Musik und Gesellschaft

Series Editors

Sarah Chaker, Institut für Musiksoziologie, Universität für Musik und darstellende Kunst, Wien, Austria

Michael Huber, Institut für Musiksoziologie, Universität für Musik und darstellende Kunst, Wien, Austria

Die Schriftenreihe „Musik und Gesellschaft", ursprünglich von Kurt Blaukopf gegründet, wird seit über 50 Jahren vom Institut für Musiksoziologie in Wien herausgegeben und seit 2017 vom Verlag Springer VS verlegt. Aktuelle Befunde musik- und kultursoziologischer Forschung sichtbar zu machen stellt das zentrale Anliegen der Reihenherausgeber*innen dar.

Reihenherausgeber*innen:
Heft 1 bis Heft 21: Kurt Blaukopf
Heft 22 bis Band 27: Irmgard Bontinck
Band 28 bis Band 37: Alfred Smudits
Ab Band 38: Sarah Chaker und Michael Huber
Aktuelle Publikationen in der Reihe „Musik und Gesellschaft":

- Smudits, Alfred (2018): Roads to Music Sociology. Reihe *Musik und Gesellschaft*, Band 37. Wiesbaden: Springer VS.
- Huber, Michael (2018): Musikhören im Zeitalter Web 2.0. Theoretische Grundlagen und empirische Befunde. Reihe *Musik und Gesellschaft*, Band 36. Wiesbaden: Springer VS.
- Niederauer, Martin (2014): Die Widerständigkeiten des Jazz. Sozialgeschichte und Improvisation unter den Imperativen der Kulturindustrie. Reihe *Musik und Gesellschaft*, Band 35. Frankfurt/Main u. a.: Peter Lang.
- Gebesmair, Andreas / Brunner, Anja / Sperlich, Regina (2014): Balkanboom! Eine Geschichte der Balkanmusik in Österreich. Reihe *Musik und Gesellschaft*, Band 34. Frankfurt/Main u. a.: Peter Lang.
- Binas-Preisendörfer, Susanne / Unseld, Melanie (Hg.)(2012): Transkulturalität und Musikvermittlung. Möglichkeiten und Herausforderungen in Forschung, Kulturpolitik und musikpädagogischer Praxis. Reihe *Musik und Gesellschaft*, Band 33. Frankfurt/Main u. a.: Peter Lang.

Weitere Informationen zur Reihe finden sich auf www.musiksoziologie.at

More information about this series at http://www.springer.com/series/15551

Andy Battentier

A Sociology of Sound Technicians

Making the Show Go on

 Springer VS

Andy Battentier
Grenoble, France

ISSN 0259-076X   ISSN 2522-8331 (electronic)
Musik und Gesellschaft
ISBN 978-3-658-33028-6   ISBN 978-3-658-33029-3 (eBook)
https://doi.org/10.1007/978-3-658-33029-3

Responsible Editor: Stefanie Eggert
This Springer VS imprint is published by the registered company Springer Fachmedien Wiesbaden
GmbH part of Springer Nature.
The registered company address is: Abraham-Lincoln-Str. 46, 65189 Wiesbaden, Germany

To Jérémy Leduc

Friend, musician, and philosopher

Who took care of everyone up to the end

Introduction: Sound Engineers as a Research Object

Now the seats are all empty

Let the roadies take the stage

Pack it up and tear it down

They're the first to come and last to leave

Jackson Browne – The Load Out (song)

In modern-day concerts it is common for musicians to publicly thank the road crew or the sound engineer. These acknowledgements often say that it would not have been possible for the concert to take place without these background workers setting up the stage, lights and sound. Some artists go even further, and write songs paying tribute to the road crews. One such is Jackson Browne's The Load Out, which is about the tasks of the road crew and life on the road. The end of the song emphasizes the dependency of everyone involved in the show; musicians, audiences and promoters, has on the roadies:

People stay just a little bit longer

We want to play -- just a little bit longer

Now the promoter don't mind

And the union don't mind

If we take a little time

And we leave it all behind and sing

One more song –

Oh, won't you stay just a little bit longer

Please, please, please, say you will

Say you will"

Jackson Browne—The Load Out (song)

In another style, the parody rock band Tenacious D also mentions the importance of the road crew for making the show happen:

"Well it's 3pm, time to lug the gear

Gotta get it on the stage

My muscles flex, my fuckin' sweat will save the day [...]

And when the crowd roars

Brings a tear drop to the roadie's eyes

Tears of pride

Because he brought you the show

But you will never know

He's changing the strings

While hiding in the wings

No matter how hard, the show must go on"

Tenacious D—Roadie (song)

Those tributes are sometimes personally addressed, reflecting the closeness of some musicians to their road crews. For example, *Alan's Psychedelic Breakfast*, on Pink Floyd's album "Atom Heart Mother", refers to their producer and sound engineer Alan Parsons. *One Tree Hill* of U2 pays tribute to Greg Caroll, a sound technician friend of the band, who died in a motorcycle accident. Neil Young's *Tonight's the Night* describes the way the folk singer was shaken by the death due to overdose of one of his road crew members:

Bruce Berry was a working man

He used to load that Econoline van.

A sparkle was in his eye

But his life was in his hands. [...]

Early in the mornin'

at the break of day

He used to sleep

until the afternoon.

If you never heard him sing

I guess you won't too soon.

guess you won't too soon.

'Cause people let me tell you

It sent a chill

up and down my spine

When I picked up the telephone

And heard that he'd died

out on the mainline.

Neil Young—Tonight's the night (song)

In all these songs, road crews i.e. people handling technical matters appear both as essential to the show, emotionally involved in it, and often close to the artists. Since Howard Becker (1982), we know that artistic production is a collective process, not simply emerging from the creativity of specific individuals. A whole chain of interactions is necessary to make an art object exist *as it is*. Materials upon which artistic works are performed and through which they are mediated must be produced and provided: film and cameras for photography and cinema, musical instruments for musicians. Distribution chains have to make artistic works available to audiences, while critics, media, and commentators provide visibility to certain works and to guide audiences' consumption. Becker shows that each element of an art world influences the final shape that an artwork will take, beyond the creative impulse given by artists.

Despite this emphasis on the collective dimension of artistic works, in Becker's accounts, technicians fall into the category of "support personnel", which is loosely described as "a miscellaneous category designed to hold whatever the other categories do not make an easy place for" (Becker, 1982, p. 2). Becker only sees support personnel as a source of people "assisting the artist" (Becker, 1982, p. 77), "hired for [their] ability to perform one function" (Becker, 1982, p. 82) and "interchangeable" (Becker, 1982, p. 81). They do not make "choices that give the work its artistic importance and integrity" (Becker, 1982, p. 77). In sum, technicians' contribution to the artistic production process is construed as limited to the performance of delegated tasks, remote from the production of meaning.

This claim has been poorly empirically investigated. Indeed, technicians have rarely been the direct focus of academic thinking. However, the few empirical works focused on the technicians' role in artistic production tend to draw a different picture. Kealy (1979) has shown that rock music emerged in the US partly because of the appearance of new forms of collaboration between musicians and technicians in recording studios: the contribution of so-called "support personnel" is shown here as critical to the emergence of an artistic movement. Hennion (1981) showed how pop artists can be driven by art directors* in collaboration with sound technicians during recording sessions, notably in order to meet certain standards in terms of sound characteristics. They guide musicians on how they should position themselves in order to sound good, they give feedback on how the record sounds, and they advise them on how to make their music respond positively to the environment of a recording studio. They perform, in this last regard, a work of "care", helping musicians to play from the heart in an intimidating environment (Leyshon, 2009). Technicians also take a significant number of aesthetic decisions in the ethnographic accounts of recording sessions produced by Rudent (2008) and Perrenoud (2007). In the former example, the technician proposed by the musician's label will become the art director* of the record and have an important influence on the structure of songs, which he will negotiate with the lead singer. In the latter, musicians fairly inexperienced in recording sessions are guided by the studio's technician, in the same way described by Hennion. The technician will also provide artistic proposals, although s/he is *a priori* only engaged to record what the musicians create. Finally, outside of the realm of music, Domínguez Rubio (2012) has shown that Spiral Jetty, an artwork of landscape sculpture in the US, couldn't have emerged from Salt Lake without the input of those he calls "the crew", i.e. the construction workers in charge of providing technical answers to the artist's aesthetic projections. In all these articles, we can see that technicians make very concrete aesthetic contributions during the production of an artwork, which contradicts the idea that there are no "choices that give the work its artistic importance and integrity" in the work of "support personnel" (Becker, 1982).

Conceptualizing Technicians as Technical Intermediaries

Hence, there is a gap between the way a technicians' role in art worlds is theoretically conceptualized, and the empirical accounts of their contribution within them. In fact, the notion of "support personnel" might be problematic in itself. As Becker notes, the convention of dividing artistic production into a core activity performed by artists, and support personnel doing the less relevant and less talent-requiring tasks, is endogenous to art worlds rather than driven by sociological analysis (Becker, 1982, p. 77). The criteria to separate an "artist" from the "support personnel" are unclear: "every function in an art world can be taken seriously as art, and everything that even the most accepted artist does can become support work for someone else" (Becker, 1982, p. 91). From the perspective of a director or a composer, the performer of the violin part of a symphony may be considered to be "supporting" (Becker, 1982, p. 80–83). Hence, "support personnel" does not precisely capture, theoretically or empirically, a category of actors with a specific role. Instead, it describes a system of delegation of tasks, where tasks considered "not artistic" are delegated to a range of replaceable people. This system can take many different forms and is not fixed in time. It therefore cannot be used as a category describing the group of technicians we are interested in.

The distribution of tasks in processes of cultural production has been studied in recent years through the perspective of "cultural intermediaries" (Lizé, 2016; Maguire & Matthews, 2012; Negus, 2002). This term refers to a group of people who in recent decades have become much more salient in cultural production: the agents*, bookers*, public relations persons, buyers and sellers of cultural products. In cultural production, such cultural intermediaries "construct value, by framing how others—end consumers, as well as other market actors including other cultural intermediaries—engage with goods, affecting and effecting others' orientations towards those goods as legitimate—with 'goods' understood to include material products as well as services, ideas and behaviors" (Maguire & Matthews, 2012, p. 552). Cultural intermediaries "impact upon notions of what, and thereby who is legitimate, desirable and worthy, and thus by definition what and who is not" (Maguire & Matthews, 2012, p. 552).

Can technicians be considered as "cultural intermediaries"? Technical crews do engage in a process of mediation: they materialize artistic ideas into specific objects, and thus directly impact the way viewers, listeners, readers etc. will engage with cultural goods. However, they hardly fit the general understanding of what cultural intermediaries do. This becomes evident when considering Lizé's (2016) typology of cultural intermediaries. All these intermediaries focus for instance on market access of artworks and artists, by bringing symbolic goods to the attention of the general public as directors of institutions, mediators, or distributors. They frame the opinion of audiences by sharing their own opinions as critics. They shape and produce symbolic capital. Moreover, they may even create encounters in order to trigger the production of a fruitful artistic project (Lingo & O'Mahony, 2010). In doing so, they create the cultural framework in which symbolic goods are seen as legitimate, meaningful, or worthy of interest. However, what they do not do is directly modify the material shape of artworks. This work is handled by tech crews: cameramen, editors, stagehands*[1], or sound engineers. And this group is *de facto* excluded from "cultural mediation", as their role does not grant them the power to create market access. This particular mode of exclusion in cultural production has been noted by Hesmondhalgh (2006), who called for a more inclusive framework under the label of "project team". Wright (2005) aimed to capture this distinction by separating the "makers of meaning" from the "makers of things". The term "makers of meaning" might be misleading, as makers of things do produce meaning as well. But there is indeed a clear distinction between the work of cultural intermediaries, who work on symbolic capital (Dubois, Durand, & Winkin, 2013) of cultural productions, and the work of technicians who work on their material shape.

I will conceptually denote this difference by framing the "makers of things" that are technicians as a category in itself, that I will call "technical intermediaries". Like cultural intermediaries, they are involved in mediation, but their mediation is of a different nature. Technical intermediaries forge a link between the artistic work, as created by artists, and the audience. They do so by transforming the material shape of artworks. In contrast to cultural intermediaries, they do not directly work on symbolic capital—in the sense of legitimacy, framing, or "buzz"—but they add something material. Hence, cultural and technical intermediaries work in a complementary way. The former attempt to place the cultural object in a larger cultural and economic framework in order for it to reach an audience, but without directly modifying the material content. The latter, on the other hand, work directly on material properties to ensure that the artwork will match both the artist's aesthetic views, the cultural intermediaries' needs—and ideally, the audience's taste buds.

---

[1] All words with a * are defined in the Glossary p. 130

A Focus on Humans

The notion of technical intermediaries places the analysis' focus on technicians themselves, understood as the people working on the materialization of artistic ideas embedded in the semantic and economic frameworks designed by cultural intermediaries. Therefore, it has to be clear to the reader that this dissertation's subject is not sound in itself, but the people who are in charge of shaping it.

This analytic standpoint marks a crucial difference from a lot of the existing literature related to this topic, which generally starts with a reflection on the medium of sound and discusses the changes brought about by developments related to sound-reproducing technologies. Some works propose a historical approach: Maisonneuve (2009) interestingly shows how the widespread adoption of the phonograph changed music consumption in the first half of the 20th century, laying the foundations of contemporary consumption. On the production side, Horning (2013) browses the history of recording studios in the US in the same period, showing that while recorded music progressively took hold of a house's rooms, music-lovers ears and public spaces, a whole range of creative practices were developed in recording studios. This led to a reinvention of composition practices, and progressive growth of the centrality of studio work in musicians' careers. Therefore, producers and engineers became more and more central in musical creative processes, both as active builders of the record's aesthetic (Hennion, 1981; Horning, 2013; Kealy, 1979) and as assistants to musicians, guiding them towards the desired result. For instance, Leyshon (2009) and Watson (2014) both emphasize the "emotional labour" performed by sound engineers, consisting of fostering the emotional state in which a musician is able to play at his/her best despite the sterile atmosphere of the studio.

These studies efficiently explore the consequences of the use of sound reproduction technologies on music itself and on the practices related to it. However, the fact that they start their analysis from "sound"—and often the studio itself, as a place or a technology—limits their scope. It is significant that, to my surprise, I have been unable to find works using similar methods to study how sound is treated in concerts. This focus on studio work is paradoxical, especially for the works focusing on sound-reproducing technology. Indeed, the process through which music is recorded or amplified is notably similar, and its use is just as systematic for concerts and for records. By leaving aside live performances, these studies treat music as something produced to be carved on a material, reproducible object; or conveyed through space by the medium of the radio. They ignore the fact that music is, first and foremost, a performance during which a player and a listener are connected by a common language, and that recorded music is one way among others to realize this connection.

This is why, in this dissertation, I opted for a performance-oriented approach to sound, focusing on the people who make sound, rather than sound itself or the tools and places through which it is done. I will show that these tools are integrated into a process of cultural production which can be understood through the analysis of human relationships. I hope to convince the reader that such an approach constitutes a better way to understand how technology is integrated in music practices, how it transforms it and therefore how it transforms the heard result.

The Contribution of Sound Engineers to Music

In the following pages, I will therefore extensively account for the contribution of technical intermediaries' actions to the sound of musical performances. I will focus on people that set up, manipulate, and dismantle the machines brought into the process of musical production for the purpose of sound reproduction. This group, in Western Europe at least, tends to be systematically present in musical performances. Indeed, music is an art world in which the contribution of sound engineers (or technicians), as technical intermediaries handling sound reproducing technologies, has been essential for more than a century. It is interesting to note that after the invention of the phonograph by Thomas Edison in 1877 (the original purpose of which was to ease the transcription of speeches and was not thought of for musical use), the market for recorded music became run by a small number of global players within a couple of decades, and has remained so until today (Tournès, 2008). The large-scale adoption of domestic music-reading devices drastically changed the ways music was consumed, even changing the conventions of music worlds: while one of the selling points of records before World War II was their similarity to live concerts, fulfilling the desire to have an "orchestra at home" (Maisonneuve, 2009, p. 59), this convention reversed in the sixties in the US, as studio sound engineers started to be brought on tour by bands in order to reproduce on stage the sound designed in the studio (Kealy, 1979). Live concerts themselves are heavily influenced by the use of sound reproduction devices. To take just one example, Queen's performance of Bohemian Rhapsody in Wembley Stadium, recently glorified in the biopic of the same name (Singer, 2018), would have looked pretty different without a sound system carrying music to the ears of the estimated 72,000 people attending. However, the issue of sound amplification is not only a concern for gigantic performances. "Ordinary musicians" from local bands also have to deal with their share of technical issues, potentially by embodying both the roles of artists and technical intermediaries (Perrenoud, 2007). However, music can be present in other forms of cultural production, and sound engineers tend to be present in these cases as well. Theatrical performances also extensively employ sound reproduction devices for playing recorded or live music. Finally, the use of a sound reproduction device is by nature systematically implied in mediatized performances, whether through radio, television or film.

Hence, sound technicians handling the reproduction of music through speakers are systematically found wherever music is present. Focusing on them is therefore a way to study the contribution of technical intermediaries in a very large variety of art worlds, in different institutional and aesthetic contexts. However, the essence of their work can be summarized in a way that does not vary with context, giving us the ability to make relevant comparisons. Indeed, music reproduction through sound devices takes two forms: recording and amplification. Recording transfers this sound onto a medium and allows it to be played in a different time and place, while amplification restitutes* quasi-simultaneously the sound on speakers at an increased volume in the context of a live performance. Both processes are realized with the use of a series of tools. In a nutshell, microphones capture* the sound of acoustic sources and transform variations in air pressure into an electrical signal. This signal is transferred through cables to a workstation composed of a mixing desk and audio treatment devices. In the case of recording, the signal to be worked on is written to a medium that will be played by various computers, stereo systems, or portable devices. In the case of amplification, the signal will be directly transferred to amplifiers and restituted* live on speakers. The whole process is called the "audio chain" (Figure 1), and sound engineers are the technical intermediaries responsible for its smooth operation (Mercier (dir.), 2017).

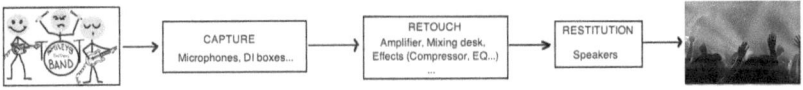

Figure 1   The audio chain.
-picture on the right by Edward Cisneros, on the left by Dunive, montage by the author

The output from the audio chain, which can be physically described as variations of acoustic pressure, will travel to the ears of audience members, and will be identified as music. Sound engineers are thus a proxy through which a symbolic good, in this case music, passes in order to be materially intelligible to the people to which it is addressed, in this case an audience. In other words, the technicians shape the object that will be the mutual focus of attention in musical performances. The level of intensity of this focus is related to the intensity of emotional effects that such events intend to trigger. Indeed, musical performances can be framed as interaction rituals (Collins, 2004), in which people engage in order to be involved in collective emotional entrainment. Such entrainment is characteristic of "re-fused" performances (Alexander, 2004), which produce the effects of a successful interaction ritual, where participants leave with increased emotional energy, membership symbols and group solidarity. Hence, by focusing on how sound technicians craft an object of collective focus, we can better understand how emotional experiences typical of successful performances are planned, built, and grounded in materials and environments. It explains how a meaningful system of symbols is incarnated in a material object, how this incarnation is perceived, and how it potentially manages to trigger emotional escalation. In both Alexander and Collins approaches, it is possible to precisely locate the moment and nature of sound technicians' intervention in order to study it in detail.

Collins' "interaction rituals" are simply an interaction between at least two human beings, realized with the influence of a specific cultural context, which can strengthen or weaken the bonds between people involved in this interaction. Its principle consists in physically gathering two or more people, who mutually identify as members of the same group and who engage in a mutual focus of attention. This attention is likely to grow if they receive positive feedback from fellow participants, building emotional entrainment and can potentially produce collective effervescence (Durkheim, 1912), even though it is extremely rare that interaction rituals lead to the form of religious trance described by Durkheim. The notion of interaction ritual, in sum, covers pretty much any human interaction. Collins illustrated his theory with examples such as 9/11, a running race, sexual interaction or tobacco rituals. He later applied it to violence (Collins, 2013), and other authors have explored interaction ritual chains in sports fandom (Cottingham, 2012), social movements (Liebst, 2019), or choral music (Heider & Warner, 2010). Collins' approach provides a comprehensive model of how human interaction fosters social bonds, and can thus be applied for events of various levels of intensity, size, and stratification. If we view musical performances as interaction rituals, sound engineers act on the stage-setting moment which precedes the performances. Rituals need to be situated in material places, and setting them up requires the mobilization a number of resources including technical intermediaries who are part of the personnel. Collins claims that the enthusiasm for gathering these resources is dependent on the importance of the benefits in emotional energy hoped for by the ritual's participants (Collins, 2004, p. 141–182). Through Collins' eyes then, sound engineers are people using resources gathered to materialize the interaction rituals which are musical performances, the quantity of resources is related to the level of emotional entrainment foreseen by the performances' participants.

While Collins' approach accounts for almost all human interactions, Alexander's perspective is more focused on *moments of gathering* in complex and stratified societies. Starting from Durkheim's works and extensively drawing on Turner (Turner & Schechner, 1995; Turner, 1982) and Schechner (Schechner, 1974, 1985; Schechner & Schuman, 1976) works on the boundaries between rituals and theatre, he developed a theory of performances with the explicit purpose of demonstrating that culture is an autonomous object of social life, producing social dynamics on its own, rather than determined by external socio-economic factors (Roberge, 2009). Alexander assumes that as societies become larger, more diverse and stratified, a group of performance specialists emerge, who take charge of what happens on stage and produce something to be shown to another group of non-specialists. Performances can be artistic, such as a theatrical play or a concert, but they are not limited to that: political rallies are also cultural performances. Performances are scripted: the specialists pick and assemble elements from a larger cultural background, in which all participants are included, in order to draft a scenario and an intended meaning for the event. The larger, more stratified and diverse the group of participants of the performance, the harder it is to create a script that would be able to affect the whole audience. This phenomenon is called "de-fusion": the fact that before the performance, participants are members of separate social groups. The objective of cultural performances is thus to "re-fuse", which means to achieve a form of coherence between script and acting, leading the participants to a form of emotional convergence, incrementing meanings and ties among them. If a performance "re-fuses", it becomes "ritual-like", meaning that it produces the outcomes of a successful interaction ritual: membership symbols, group solidarity and emotional energy. In Alexander's perspective, sound engineers intervene at the moment of *mise-en-scène*, when the script designed by artists and cultural intermediaries is confronted with acting and the material environment in which it is performed, allowing it to "walk and talk" (Alexander, 2004, p. 554) and become a meaningful cultural performance. Sound engineers therefore must pave the way, during the *mise-en-scène*, for the script to be meaningfully acted in concrete material conditions.

These theoretical frameworks, centred on the moments and meanings of musical performances, are thus the most appropriate ways to understand how technical intermediaries' actions contribute to the production of meaningful events. From a performance perspective, the distinction between production and consumption is blurred: all the people involved (musicians, cultural intermediaries, technical intermediaries *and* audiences) contribute to the achievement of emotional spin-up (McCormick, 2006). Under such a theoretical gaze, technical intermediaries manage the adaptation of the performance's script to the practicalities of space, time and contingencies. These practicalities can thwart the process of a musical performance, and technicians are there to make sure that there is a smooth implementation of the script, and thus re-fusion and collective emotional entrainment. Hence, from a performance perspective, sound-reproducing technologies are tools in the hands of people serving the purposes of producing ritual-like outcomes in musical performances. I intend, in the following pages, to analyse the different ways through which technicians, in the shadows of backstage, contribute to building the ritual-like outcomes of musical performances.

Method: A Comparative Approach of Sound Technicians in France and the Netherlands

If all interaction rituals have the potential to provoke emotional entrainment, leading to emotional energy, group solidarity and membership symbols, the way in which its realization manifests itself depends on the particular script of the performance, and therefore on the cultural background of participants. Hence, in order to understand how technicians contribute to the production of emotional entrainment, we need to understand what is expected from them in various national, professional, and artistic contexts.

A Binary Comparison Between France and The Netherlands

Despite the emergence of transnational fields of cultural production (Franssen & Kuipers, 2013; Kuipers, 2011), national boundaries remain an important factor influencing both cultural production and reception. Transnational popular culture can be construed as an element of cultural distinction by the national dominant classes (Hedegard, 2015), national cultural practices reach transnational fields once they have passed through nationally situated cultural centres (Jiménez Sedano, 2019), and transnational cultural objects are modified in order to be broadcast in a new nation (Kaplan, 2012; Kuipers, 2015). Given the constant influence of national factors in explaining cultural phenomena, I expect that national contexts will influence the way technicians contribute to the realization of musical performances.

In order to account for this influence, I will perform a binary comparison (Dogan, 2008) between France and The Netherlands. These two countries' musical fields have many similarities. Despite French state centralism and the Dutch federalist vision of nationhood, both benefited from national cultural policies in the second half of the 20th century, which were both oriented towards the subsidization of high arts, namely classical music and opera. These subsidies were justified in both countries by the necessity of cultural reconstruction in the aftermath of the war, and to counter the influence of mass culture imported from the US. Their general purpose was to produce and disseminate national highbrow culture among the population (Poirrier, 2002; Hamersveld, 2009). At the beginning of the seventies, popular music started to gain recognition in both countries and benefited from cultural policies as well, although the subsidization of highbrow music remained the focus of resources. Since the 1980s and the pressure on state budgets resulting from the economic crisis, both states slowed and eventually stopped the expansion of their funding, although the Netherlands have arguably gone further in this dynamic, especially in recent years. The government voted for heavy cuts in the cultural policies budget in 2011, publicly embraced new public management in cultural institutions and began to "redress the addiction of the [cultural] sector to subsidies" (Meerkerk & Hoogen, 2018). Recent years have seen a tendency to promote "cultural entrepreneurship" in The Netherlands, while France generally held onto its historical concept of culture as a "public service".

Music venues and initiatives are spread across both countries, but the economic means and principal institutions are heavily concentrated in their capitals, Paris and Amsterdam (Brandellero & Pfeffer, 2015; Guibert, 2006). As a consequence, technical intermediaries making a living working in musical fields tend to concentrate in these cities. However, the level of commodification of the two national labour markets is very different. Artists and technical intermediaries tend to work on short-term contracts rather than being permanently employed in a single organisation (Menger, 1999). In the Netherlands, these freelancers are highly individually responsible for ensuring their financing of health and unemployment risks. They fall under the regime of *Zelfstandige Zonder Personeel* (independent with no staff), and can potentially be employed on 0-hour contracts, which does bring health coverage but does not guarantee regular income (Vonk & Jansen, 2017). In France, an *ad hoc* regime of unemployment insurance specifically impacts freelancers in the performing arts. In principle, short-term contracts are salaried contracts, and not a contract between an independent entrepreneur and a client. As a result, freelancers in France are covered by health insurance.

Moreover, once a freelancer has worked more than 507 hours within a year they then benefit from unemployment welfare during the periods in which they are not employed (Menger, 2015). This policy intends to provide features of a stable employment relationship (Bosch, 2004) to freelancers in the performing arts, while preserving the requirement of the sector to recruit on short-term bases.

Hence, both musical fields are similar in their national historical cultural policies, although these policies were justified by slightly different rationales due to the differences in how these states perceive state centralization. Furthermore, quite divergent directions have been taken during the last decade, where public funding underwent a significant reduction in the Netherlands. Both fields are marked by a tension between the promotion of local, national productions supported by national cultural policies and the influence of transnational culture. Finally, technical intermediaries in France act in a labour market that is much more decommodified. This decommodification has been shown to be a central factor for determining the way artists and technicians manage their time spent on creative activities (Corsani, 2012; Grégoire, 2012, 2013).

Institutional Comparison

Technicians' labour has a cost, paid for by music venues organizing musical performances. Both the cost of technicians' labour and the ability of venues to afford it are directly influenced by the differences between French and Dutch national policies. The relative social security from which French technicians' benefit, and their status as salaried workers, are likely to make their workforce more costly for venues than in the Netherlands, where technicians have less social security. Besides, given the assumed market orientation of recent cultural policies, Dutch venues are likely to rely more extensively on autonomous income, whereas French venues can benefit from more public funding. These elements shape the "material investment in costs of ritual production" described by Collins (2004, p. 141–182). The labour of technicians is a necessary condition for a performance's realization. However, the income that technicians earn from their work, both through their net income and the social benefits attached to their contributions (health and unemployment insurance), depends on the economic model of venues funding their work. This income influences their ability to make a living from their technical activities, and therefore shapes their careers, the type of collaboration they will have with artists, and therefore their contribution to emotional entrainment in the performances in which they work.

In order to answer the central question of this thesis, it is necessary to question the effects of the institutional contexts in which technicians' labour is embedded. For this part of the analysis, I rely on the data that has been made available respectively by the *Fédération des Lieux de Musique Actuelles* (FEDELIMA) and the *Vereniging Nederlands Poppodia en Festivals* (VNPF). Both organizations have a similar role in their respective countries: they are national associations of live music venues. They are part of the *Live DMA*, a European-wide network of similar organizations. Both organizations pool information and coordinate resources in the live pop music sector. The venues affiliated with their networks can be for-profit or non-profit. They have a capacity ranging from a few dozens to a few thousand people. They program local, national and international artists, and are therefore connected to national and transnational culture. These two networks cover a wide scope and proportion of their respective countries' venues, although publicly-funded classical music venues, arenas receiving pop superstars, and small local venues are not among their affiliates.

Both networks receive institutional recognition, as they provide statistical data for their respective ministries of culture in the field of pop music. One of the ways they do so is through an annual survey of their members since 1999 for FEDELIMA, and since 2003 for VNPF. In this survey, music venues provide detailed information on their activities, their income, their overhead, and their workforce. I have been granted access to the detailed answers of the survey under the condition to respect the anonymity of music venues, and I will use them to study the effect of national policies on the institutional contexts in which sound technicians work.

Comparison of Technicians' Professional Trajectories

Within the economic conditions of their employment, how are sound technicians integrated in the art worlds in which they work? Who hires them? What kinds of tasks do they perform? To what extent are they involved in artistic matters? What is the shape of their professional lives? Are they restricted to an artistic form, a specific place, a particular technology, an initial diploma? The answers to these questions will allow to draw a picture of the dynamics of technicians' careers, a domain completely ignored by the current literature. It will help us to understand how art worlds, understood as a collective of artists, cultural and technical inter-mediaries, and audiences who aiming to produce emotional entrainment during a music performance, are formed.

In order to answer these questions, I have collected information on technicians' professional trajectories through semi-directed interviews. The interviews
aimed to gather information about their social trajectories before and during their
career, the current state of their career, and their professional practices and ethos.
After a general review of these issues, I asked for a more detailed review of their
different jobs in the last 12 months, in order to get more precise and fundamental
information. The interviews lasted between forty-five minutes and two and a half
hours. The interview grid is presented in Appendix B.

Two rounds of ethnographic fieldwork have been done for this study. The first
one happened in France between December 2015 and June 2016. In this period, I
conducted 17 interviews with sound technicians working at least partly in music
(live or recorded). Sixteen were living in Paris and its suburbs, one in the region
of Lyon (France). The second fieldwork round was done between January 2018
and June 2018 in the Netherlands. I interviewed 11 sound engineers working at
least partly in music. Most of them were living in the Amsterdam area, one of
them was from Utrecht and another one from Alkmaar. In general, respondents
in both countries worked in different places throughout their respective countries,
occasionally internationally. Additionally, I took the opportunity to interview a
Canadian respondent, living in Montréal, as he worked regularly on the tours of
globally-known artists and shows. Respondents were recruited either through personal or professional contacts, or through messages posted on Facebook groups
for communities of musicians or technicians, with the precondition that they were
not known to me prior to the interview. Snowball sampling has also been applied,
as the respondents recruited first sometimes offered to introduce me to other technicians. Furthermore, when I was granted the opportunity to follow a respondent
during his/her work, I tried to get in contact with other technicians in that loca-
tion, who were sometimes unknown to the respondent I initially followed. For
instance, the front of house sound engineer can tour with a band and meet the
venue's sound engineer for the first time. If I was accompanying one of them,
I systematically suggested an interview to the second one. Table 1 presents characteristics of the interviewed panel, and an anonymized list of respondents is
presented in Appendix A. All respondents in the French field are French, while 9
respondents out of 11 in the Dutch field are Dutch (Table 1).

Comparison of Working Practices in Different Artistic Contexts

After showing how national contexts and professional networks constrain the
contribution of technicians to musical performances, I will focus on how this

Table 1 Age, Gender, Education distribution of respondents

	France	The Netherlands	Total
Age			
20–29	3	5	8
30–39	12	3	15
40–49	2	1	3
50+	0	2	2
Total	17	11	28
Gender			
Female	3	0	3
Male	14	11	25
Total	17	11	28
Education in the field			
Learned on the job	3*	4	7
1 year of formal training	2	2**	4
2 years	4	2	6
3 years	1	3	4
5 years	7	0	7
Total	17	11	28

*Two of them have a university degree equivalent to respectively 4 and 5 years of study, but not in the field of sound engineering
**One of them has a bachelor's degree, not in the field of sound engineering

contribution varies according to the artistic ambitions and conventions of different art worlds. For this, I rely on participant observation during the preparation and execution of musical performances.

After the interview, I asked respondents whether it might be possible to be accepted as an observer at their workplaces. Eighteen of them agreed and granted me access to 30 observations of musical performances. Most of these observations were concerts, in which I was either following the sound engineer of the invited band, or the one employed by the venue, in charge of welcoming the band and their assigned technician. These concerts were either part of a tour, or included in a festival*, or single events that were not planned to be repeated. A handful of

observations were theatrical plays where music had an important role. Incidentally, I observed one recording session, one rehearsal, one live radio show and one
dance contest. Additionally, I worked for a week as a volunteer aide in a festival*
of world music. In sum, I have observed a wide variety of musical events in terms
of musical genre, range and economic model, with musicians of local, national
and international reputation. The observations are summarized in Table 2.

Access was helped by the fact that I have worked as a self-taught sound technician before studying social sciences. I started as the 'sound guy' of my own
band and friends' bands in my high school years, while doing internships in local
theatres to learn about the profession. I continued in my first years of university
and started to work on various projects: I did the sound design of several plays
for student and semi-professional theatre companies, got involved in live concerts
and festivals*, either as an aide, a sound engineer or a technical director*. I was
however split between my interest in this job and in the one sparked by the dis-
covery of social sciences in my bachelor's degree. I missed the entry exam to a
renowned sound school and worked as a sound technician for about 8 months.
After that, my interest in social sciences took over and I started a master's degree
in order to do research. As a consequence of these experiences, I am familiar with
sound technicians' working conditions, I have a practical knowledge of the secrets
of the profession, of the rhythm of a typical day in different working contexts,
and I can even work as a sound technician occasionally.

This broad range of contexts might be puzzling at first glance. This set of
observations includes large and small concerts, classical music and cover bands.
But art worlds follow different dividing lines from the perspective of technicians.
It is common for the same person to work across genres, artistic forms, and scales.
It is thus necessary, to understand where these dividing lines are in order to form a
corpus of data unrestricted by conceptions that apply to other actors in art worlds,
such as artists or cultural intermediaries. I applied a systematic approach to all
these observations, except for the one in which I was employed.

The observation consisted of a timeline recording of the process of *mise-
en-scène*, up to the end of the show and sometimes including its dismantling.
In an observation method similar to the one adopted by Bechky (2006), I was
observing actions undertaken by the participants of the concert, as well as their
different facial expressions and verbal exchanges. In contrast to Bechky though,
I did not have a role in the process of production but assumed the position of an
observer: I was being as discrete as possible, taking notes in a notebook in the
background. Being forgotten was relatively easy as the *mise-en-scène* is a moment
of intense focus for the actors involved, venues generally require darkness, and

Table 2 Characteristics of observations

Ref Obs	N01	N02	N03	N04	N05	N06	N07	N08	N09	N10
Country	Netherlands	Netherlands	Netherlands	Netherlands	Netherlands	Netherlands	Netherlands	Netherlands	Netherlands	Netherlands
Performance type	Concert	Concert	Concert	Radio	Concert	Recording	Concert	Concert	Concert	Concert
Style	Rock	Country	Pop	Folk	Pop	Pop/Hip-Hop	Covers	EDM	EDM	Fusion
Number ART	8	15	7	1	14	2	80	4	1	6
Number AUD	70	450	1500	NC	200	NP	250	500	1350	30
Number CI*	3	1	3	1	4	0	3	1	2	1
Number TI**	3	6	7	1	4	1	2	2	2	1
Interior/Exterior setting	Int	Int	Int	Int	Int	Int	Int	Int	Int	Int
Origin of the artist	Int/Loc	International	Int/Loc	Local	Local	Local	Local	International	Unknown	Local
Economic model	Non Lucrative	Lucrative	Lucrative	Non Lucrative	Lucrative	Lucrative	Non Lucrative	Lucrative	Lucrative	Non Lucrative

Ref Obs	F01	F02	F03	F04	F05	F06	F07	F08	F09	F10
Country	France	France	France	France	France	France	France	France	France	France
Performance type	Rehearsal	Dance contest	Concert	Concert	Concert	Concert	Concert	Concert	Concert	Theater
Style	Electro	Hip Hop	Electro	Pop/Soul	Pop/Rock	Pop	Trip-Hop	Opera	Chanson	Contemporary
Number ART	8	100 + 6	4	6	7	9	7	200	14	24
Number AUD	NP	500	700	500	400	80	50	2400	500	300
Number CI	1	1	3	2	2	3	2	0	1	3
Number TI	6	2	4	5	12	2	4	6	10	7

(continued)

Table 2 (continued)

Ref Obs	F01	F02	F03	F04	F05	F06	F07	F08	F09	F10
Interior/Exterior setting	Int	Int	Int	Int	Int	Int	Int	Int	Int	Int
Origin of the artist	National	International	Internatio-nal	International	Internatio-nal	Local	Local	International	National	National
Economic model	Lucrative	Lucrative	Lucrative	Lucrative	Lucrative	Subsidized	Non Lucrative	Subsidized	Subsidized	Subsidized
Ref Obs	F11	F12	F13a	F14	F15	F16	F17	F18	F19	
Country	France	France	France	France	France	France	France	France	France	
Performance type	Theater	Theater	Concert	Concert	Concert	Concert	Concert	Concert	Concert	
Style	Instrumental	Instrumental	Variétés	Jazz	World	Mixed (Festival)	Electro	Pop/Electro	Pop/Rock	
Number ART	3	5	1	8	7	48	6	env.40	4	
Number AUD	300	200	20	30	80	100–3000	330	150–900	1000	
Number CI	1	1	2	1	0	2 + 1	1	2	0	
Number TI	5	14	1	2	7	11	6	11	6	
Interior/Exterior setting	Int	Int	Int	Int	Int	Ext	Int	Int	Int	
Origin of the artist	Local	National	Local	Local	National	Loc/Nat/Int	Local	National	National	
Economic model	Subsidized	Subsidized	Lucrative	Lucrative	Subsidized	Non Lucrative	Non Lucrative	Non Lucrative	Non Lucrative	

* Cultural intermediaries
** Technical intermediaries

actors are likely to have only minimally modified their behaviour due to my presence. However, news that some sociologist was quietly observing somewhere often spread fast, and my behaviour generally raised curiosity. I was often questioned by people taking a break about why I was taking notes. These interactions provided opportunities for informal discussions and my presence was generally welcomed after a little explanation. In the French fieldwork, I also did brief interviews with musicians when possible, which provided some information about their social and musical background, as well as the type of collaboration they had with their sound engineer (interview grid in Appendix C). However, as the musicians were not always receptive, obtaining these interviews was quite energy consuming to the detriment of focusing on the ongoing action, and I was less sure of the acceptability of the process in an unknown context, I did not proceed with these interviews in the Dutch fieldwork. The biggest difficulty during observations consisted of finding an appropriate observation point. Being too visible would have disrupted working processes and my presence would have been perceived as an inconvenience, but being too remote from the action would have made me miss some important steps of the *mise-en-scène* process. The range of acceptable positions was thus narrow and needed constant vigilance. However, I always managed to get an overview of the process, and was never expelled from an observation session.

The material drawn from these observations is thus an overview of the *mise-en-scène* moment from the perspective of technicians, at the moment in which they are central to the performance, i.e. the moment where they define the material form of the scripted object of the performance, which must meet the implicit and explicit expectations of artists, cultural intermediaries, and audiences. This corpus will be analysed with different methods in order to answer the specific questions outlined above, and these methods will be discussed in each chapter.

Thesis Layout

This dissertation is built as a funnel, each chapter questioning how sound engineers contribute to build the conditions of a musical performance re-fusion at different levels of analysis, from macro to micro.

Chapter 1 asks whether objects and environments play an active role in determining sound engineers' actions while setting up a live performance. It focuses on two main tasks found in every concert using sound-reproduction devices: (1) balancing the different sources of sound to obtain a stereo mix, and (2) setting up and tuning the sound system for the concert. Using empirical data from the

French fieldwork[2], this chapter untangles what, in the decision-making process of sound engineers, results from pressure exerted by the material environment of the performance, and what results from a demand initiated by interpersonal and social dynamics. The objective is to test if the notions of "agency of objects" and "generalized symmetry" (Callon, 1986; Law, 2009) still hold when we consider the work of technical intermediaries. The results invite us to consider objects not as active actors, but passive constraints with respect to the script constructed by human actors. This chapter shows that objects and environments constitute elements that can impede the script of a musical performance, but that these elements can be overcome depending on the technical ability of the performance's participants, which relies on qualities based on socialization: the ability of technical intermediaries to communicate with artists and cultural intermediaries, as well as their knowledge and experience of the capabilities of the tools they use.

Once I establish the prevalence of human factors over material ones in the contribution of sound engineers, I proceed in Chapter 2 to the analysis of institutional differences between the French and Dutch fieldwork, I will analyse the financial structure of pop music venues in France and in the Netherlands, and especially compare how much they rely on subsidies or external personnel. This will require the development of an analysis of the role of the freelancers in performing arts unemployment regime in maintaining the economic viability of musical performances, and therefore the possibility to stage them as interaction rituals. I will show that this regime constitutes an original approach to the problem of "cost disease" typical of performing arts shown by Baumol and Bowen (1966). While the authors mentioned subsidies as the practical answer to personnel costs getting out of hand, I will show that the regime re-establishes features of a standard employment relationship for freelancers at the expense of the sectors fostering the mechanism of the cost disease. As such, the regime can be understood as a form of compensation for the non-expanding sectors by the ones whose productivity is expanding.

Having exposed the effects of institutional differences in the context in which sound engineers evolve, Chapter 3 addresses their trajectories and networks of relationships. In this chapter, I show how technicians train and how their career path in different artistic genres, positions and contexts of work is related to their first steps in the profession. I will also show how these characteristics of their different jobs include some aesthetic involvement, or to speak like Becker

---

[2]This chapter has also been published as an article (https://www.biens-symboliques.net/438), written before and during the Dutch fieldwork. However, the conclusion would have been the same if the Dutch fieldwork had been included in the analysis.

"choices that give the work its aesthetic and artistic integrity". We will see that the differences between the two national economic contexts impact this involvement, leading to national differences in the spread of technical knowledge and technician's advancement towards positions with higher aesthetic responsibilities. Finally, this chapter will highlight an unexpected role of technical intermediaries within art worlds. They are the actors with the highest mobility between the diversity of aesthetic forms they are confronted with in a professional context. As such, they absorb many varied conventions that they can transpose from one art world to another, contributing to the circulation of ideas between different art worlds necessary to fuel innovation in the arts (Patriotta & Hirsch, 2016; Uzzi & Spiro, 2005).

Chapter 4 is focused on the distribution of the power to decide the shape of music that sound engineers work upon. Technicians are under the scrutiny and pressure of artists, cultural intermediaries and audiences who have expectations, derived from the performance's script, on what music must sound like. Furthermore, technicians themselves can have their own views on what the music they work on should resemble. I will show that the way decisions on the matter are arbitrated can be understood using the socio-cultural notion of music genre extensively developed by Lena (2012). In this chapter, I empirically apply her framework to my fieldwork. I observe how the dimensions she describes are transformed into behaviours and norms in the field. The level of fit of her framework confirms her idea that direct use of the notion *music genres* (e.g. jazz, rock, or country) in sociological analysis is problematic. Indeed, it inaccurately transforms these genres into objective notions. As a consequence, it contributes on one hand to legitimizing them in their existing form, and therefore interferes with the struggles that are supposed to be internal to the field: it valorises the practitioners who agree with the definition of the scientist. On the other hand, it fills scientific arguments with ill-defined notions, subject to the controversies and interpretations of respondents and their field. As a result, it blurs the conceptual gap between symbolic and social boundaries (Lamont & Molnár, 2002) and tends to confusingly and too easily explain the latter by the former. We will see that Lena's framework is a way to clearly trace the steps between symbolic and social boundaries.

While Chapter 4 shows how participants' behaviours and expectations in musical performances can be understood through the lens of music genres, it also shows that the latter is not sufficient to understand the mechanisms of performance's re-fusion. Chapter 5 addresses the role of the relational contingencies in technicians' teams by questioning how their group engagement (Metiu & Rothbard, 2013) during the *mise-en-scène* influences a performance's ability to re-fuse.

Group engagement is a global parameter of a working team implying a tendency to efficiently proceed to problem-solving by engaging in interactions rituals within small groups highly focused on a specific task. It is triggered by a combination of individual, relational, and collective parameters through which I will study the effect with regard to the purpose of re-fusion during concerts. In doing so, this chapter constitutes an empirical application of a performance's perspective considering audiences as producers of the performance, and questions how emotional involvement of technicians contributes to promote emotional engagement during the performance itself. It will show that while the meaning of a performance is collectively construed by all participants, audiences' group engagement ultimately rests on the artists performance. This result sets the limits of intermediaries' contribution to art worlds: once the show is on, they step back and leave the floor to artists, whose performance defines the outcome of the performance as a success or a failure.

Contents

Technical Intermediaries and the Agency of Objects: Questioning Generalized Symmetry

Introduction: Producing Music and Meaning in Material Settings

In 2018, the online medium *The Verge* tried to find out how the Super Bowl half-time show, a musical event during the final of the US football championship with more than 100 million spectators from around the world, was set up in just six minutes[1]. The article cites Patrick Baltzell who was "in charge of the audio for the last 19 Super Bowls" and also "designs and mixes for the Grammys, Oscars, and presidential inaugurations". In this interview, the journalist brought to the fore the hidden artistry behind this live sound performance. Baltzell's work, which consists both of managing the minute preparations for the precise build of the stage, and of mixing the performance's sound on the spot, requires a careful coordination of material constraints and the demands of the producer and performers. The interview mentions a plethora of elements involved in creating the final performance: the weather and the stadium acoustics, the requests of the artistic performers, the broadcasting requirements—all are cited as elements taken into consideration in order to produce the best possible performance.

This constant interplay between material constraints and interpersonal demands is typical of the technical aspect of cultural production, as occurs in sound and visual productions such as the Super Bowl. Recent approaches in

---

[1] https://www.theverge.com/2018/2/2/16961244/super-bowl-halftime-show-audio-patrick-baltzell-2018

Electronic supplementary material The online version of this chapter (https://doi.org/10.1007/978-3-658-33029-3_1) contains supplementary material, which is available to authorized users.

the sociology of culture and the arts highlighted "the agency of objects": the capacity of objects to influence social situations. Inspired by Science and Technology Studies (STS) and Actor-Network Theory (ANT), so-called *new materialist* approaches in cultural sociology have placed objects and the material environment at the centre of the analysis of cultural production. Thus, they have stressed "affordances" offered by musical artefacts and objects, and the possibility for people to "attach" themselves to things (DeNora, 2000; Gomart & Hennion, 1999; Hennion, 2007). Others have stressed the ability of objects and environments to shape decision-making in cultural production (Akrich, 1992, 1993, 2010; Domínguez Rubio, 2012, 2014; Griswold, Mangione, & McDonnell, 2013; Klett, 2014). These approaches, in turn, have been criticized for overstating the influence of technologies, objects and environments in shaping social life, and thus downplaying the role of social processes, power relations and historical developments (Elder-Vass, 2008, 2015; Lettow, 2017).

This chapter aims to follow up on this critique on the "agency of objects". It questions the "generalized symmetry" (Callon, 1986), central to ANT, that places human and non-human actors at the same level of influence. Instead, it calls for an approach to materiality that accounts for the role of objects, while considering their influence as different in kind. In order to do this, it will show that while the contemporary production of a music performance sometimes engages a massive investment of technologies and objects, human relationships and processes of meaning-making remain the central factor influencing its final shape.

For this, this chapter will untangle which aspects of technical intermediaries' tasks, decision and practices result either from interindividual or material elements in live concert situations. A sound technician's work consists of various forms of mediation. First, they mediate between symbolic goods (music) and physical matter (sound waves). Second, they mediate between different categories of cultural producers: musicians, technicians and other support personnel. Thus, they perform both "material" mediation, as well as "relational mediation". Third, they mediate between musicians and their audiences, and do so quite literally: most music performed live in Western Europe today is intended to be worked on by technicians, and without their contribution would not be appreciated, or even be understood as music.

By untangling the relational and material aspects of the work of technical intermediaries, this chapter shows the primacy of the influence of interindividual processes, classically considered as "social" in the literature. Consequently, the "agency of objects" is understood as a result of what will be called "technical ability": the capacity of groups, individuals and institutions, induced by socialization, to overcome challenges posed by objects and environments and to use them as and when required.

Material Agency in Art Worlds as the Work of Technical Intermediaries

In their work technical intermediaries have to deal with objects, which in the field typically summarized as "Actor Network Theory", are framed as social actants (Akrich, Callon, & Latour, 2006; Law, 2009). One notable and contested characteristic of this conceptualization, that is central to the theoretical architecture of ANT, is that objects, environments, humans, animals, and ideas all take part equally in a network of interactions. This "generalized symmetry" (Callon, 1986; Law, 2009) leads to a concept of the social world as a series of associations between actants, i.e. humans, non-human and even symbols, all forming constantly evolving assemblages.

This approach tends to invalidate macro-sociological analysis, stating that society is built and dismantled fully at a the micro level (Latour, 2007; Law, 2009). It has been criticized for doing so, and accused of promoting a biased ontological approach (Elder-Vass, 2015; Lettow, 2017) that minimizes the influence of institutional factors and historical relations of power, while providing explanations of the social that are, in fact, "supra-societal" (Lettow, 2017, p. 112). The notion of generalized symmetry is at the centre of this controversy, as it implies a form of fuzziness in the definition of actants' empirical abilities. For instance, Elder-Vass comments on Callon's classical formulation of generalized symmetry that highlights the symmetrical agency of humans and scallops:

"Some of Callon's other attempts to treat scallops and fishermen symmetrically are frankly bizarre [...]. For example: 'In fact, the three researchers will have to lead their longest and most difficult negotiations with the scallops' (Callon, 1986, p. 211); [...] Latour, too, adopts this sort of symmetry [...]. As a literary device, such metaphors are stimulating. As a device for provoking the recognition of a gap in conventional sociological reasoning, they are effective. As a methodological requirement for sociological work, they are thoroughly misguided. Scallops don't negotiate, represent, or betray. Motors don't become interested in projects or allow or forbid anything. [...] But scallops have different causal powers from humans, and different causal powers from motors. Scallops have the power to attach themselves to rocks or to collectors used by the researcher; they do not have the power to negotiate. Motors have the power to drive vehicles in certain conditions (but not in others); they do not have the power to be interested, to allow, to forbid." (Elder-Vass, 2008, p. 468–469)

By granting objects similar social capacities to the ones granted to humans, Actor Network Theory interestingly showed that materials do influence social life, contributing to rescuing sociological analysis from exaggerated anthropocentrism, framing society as evolving independently from its environment. However, it is also true that scallops do not negotiate. Instead, it is more accurate to say that humans are trying to influence the behaviour of the animal in a controlled environment. They try to do so without knowing much theory about its behaviour, in a trial and error process that leads ultimately to producing a scientific discourse on the scallops. In Callon's article, this process is analysed from the overarching position of its designers. In turn, the people implementing the process on the ground never appear in the analysis, which gives the impression that scientists directing the project enter into a direct dialogue with the scallops, bringing the technical options tested to obtain the mollusc's "cooperation".

This gap is particularly visible in recent attempts to integrate objects into the sociology of the arts, in particular by Domínguez Rubio (2012, 2014, 2016) who appears to be split between this notion of objects' "negotiation" and his observation of people involved in the task of handling the different resistances implied by the physics of material objects and environments. For instance, in the production of the landscape artwork Spiral Jetty (2012), "the crew" is indispensable in solving the practical problems posed by an environment rather hostile to artist Robert Smithson's creativity. The crew stewards the successful production of the artwork, despite the muddiness, salinity and isolation of the place where it was built. Although his concepts of "docility" or "unruliness" of pieces of art exposed at MoMA still frame objects as active actors in their own definition as artworks, Rubio highlights the role of conservators using an array of technologies to hide the signs of decomposition of the objects (2014). Focusing on conservators in his work on the Mona Lisa (2016), he concludes the need, following Becker (1982), and Shapin (1989), to focus on "cleaners, plumbers, mechanics or conservators, who are responsible for the critical work through which objects are sustained on a daily basis, and without whom these objects, as well as the systems of meanings and value that are woven through them, would simply collapse in front of our eyes" (Domínguez Rubio, 2016, p. 82).

In sum, Rubio recognizes that associations between objects, humans and meanings require human work. He identifies people that shape the matter in order to give it an intended meaning: builders in the case of Spiral Jetty, conservators in the case of the MoMa and the Mona Lisa. These people are, in fact, technical intermediaries: they work on the material form of an artwork conceived and funded by other people. However, following the assumption of generalized symmetry, Rubio's analysis still focuses on materials framed as active actors in the meaning production process, and therefore conceptualized the work of technical intermediaries as a response to the actions of objects, rather than as action on the objects:

> *"[Q]ua things, **artworks are constantly veering away** from the object-positions to which they are subsumed."*

> *"[C]onservators attached a prosthetic measuring device [...] that record the **daily behavior of the painting**." (F. Domínguez Rubio, 2016, p. 78, authors' emphasis)*

As a result of this focus on an object's behaviour, framed as active and almost conscious, technical intermediaries themselves are hidden from view: we do not know who these people are, what their trajectories are, or how they are integrated in their working environments. Most importantly, we do not know how the answers to these questions influence their ability to create or maintain the shape in which objects are meaningful. These are the questions asked in this chapter, which will reverse Rubio's perspective: starting from the technical intermediaries' practices rather than from the materials they are working on. In doing so, generalized symmetry is called into question. This chapter's purpose is therefore to understand whether objects still appear as having active behaviour, equivalent in influence to the one of the human beings with which they are in contact, from the point of view of technical intermediaries in charge to handle the dialogue between materials and people's intentions (Barley, 1996). In fact, it will be shown that integrating objects in a system designed by humans is not so much a matter of objects' "negotiation" but of "technical ability", which is the ability of the group of humans to handle this integration. This ability relies on education, experience, and abilities to communicate. However, the point made is not to return to the mere anthropocentrism that ANT successfully debunked. Objects and human contributions in the production of cultural meaning have a qualitatively different nature, as it will be shown empirically.

Despite arguably being a central type of actor in contemporary, technology-driven cultural production, technical intermediaries have been the object of very little research in the current literature. Although ethnographic accounts of their working practices in different art worlds do exist (Horning, 2004; Kealy, 1979; Kuipers, 2015; Le Guern, 2004; Leyshon, 2009; Perrenoud, 2007; Rudent, 2008), very little conceptual work has been done so far.

A notable exception is Barley (1996). Observing that more and more jobs fell under the naming "technicians" in the US job market, but that very little was known about what this term actually meant, he and his team conducted an extensive study of 8 technical professions: programmers, science technicians, medical technicians, computer technicians, engineering technicians, emergency medical treatment technicians, and radiological technologists. Adopting an ethnographic frame of analysis, Barley and his team spent five years shadowing people in these occupations at their workplace, trying to find common features. They found that regardless of their specific occupation, what technicians had in common was their job was to "manage [an] empirical interface" (Barley, 1996, p. 418), i.e. describing material entities in a system of meaningful symbols. From there, they can provide these symbols for interpretation by another professional (physician, scientist…). In this case, they "buffer" information for another professional. They can also interpret these symbols themselves and use this information to repair the system that is designed to be used by other people. In this case, they are described as "brokers". IT crews typically perform these brokerage tasks.

Managing the empirical interface is described as the non-relational part of technicians' work, while the other tasks of brokering or buffering are described as relational. Sound technicians also share this pattern of work: they transform the music emitted by the musicians into a series of measured electric signals, and they work on these signals to transform the physical properties of the original music. However, unlike technical professions studied by Barley, they do not provide this signal to only one category of actors, but to three: artists, cultural intermediaries and audiences. In order to address the issue of generalized symmetry, it is necessary to assess whether the relational or non-relational part takes over in their working process, or if they are evenly represented in their tasks.

Method: Shadowing Sound Engineers

As mentioned in the introduction, many concerts, regardless of their genre, place or range, require the use of a public address (PA) system. No pop musician could play effectively in a stadium without the use of amplifiers. A rock concert in a small venue would amount to nothing if the non-amplified voice was swamped by the drum and guitar amps. The PA system required is always built, managed and dismantled by sound technicians—the professional group on which we are going to focus.

Whenever a PA system is set up for a musical purpose, two tasks must systematically be performed. Mixing, i.e. balancing the sources of captured* sounds of the different instruments, is one of these tasks. However, even if the speakers are permanently installed in the venue, the system must also be calibrated*, and thus tuned for the specific music that will be played on a given day. Both these tasks were monitored systematically on each observation within the field-work. They were timed, and the actions and exchanges undertaken were noted. Whenever the workload was low, informal discussions with the technicians allowed a better understanding of what was done and why. The observations made were *work contexts* that had been previously discussed in the in-depth interviews, creating a direct connection between the interviews and the observations, providing information on the importance of, respectively, relational and non-relational tasks in these contexts. Cross-cutting these inputs generates a stable picture of the decision-taking mechanisms during these two tasks, and to identify whether these mechanisms are rather influenced by material or relational elements.

The two tasks are at first glance different in nature. Technicians are alone on stage when they set up and calibrate the PA system*, while they interact a lot with other people, in particular artists, during the mixing task. This implies the direct participation of artists, as they must often give input on whether the sound appeals to them. Hence, the next sections will focus on the influence of relational and non-relational demands in the decision-making process in a task that appears predominantly relational at first glance, and another one that appears to be non-relational.

Relational Technical Mediation: Changing Material Properties of Music

One way to interpret the work of live music sound technicians is to say that they transform music, a symbolic good, into sound, a physical object. While music is described in terms of representations, cultural references, subjectivities, identities and so on, sound is described in physical terms: units of acoustic pressure, intensities of electric signal, and other physical descriptions. These symbols become visible on a mixing desk and are handled by the technician behind it: vu-meters and screens provide measures of the physical properties of sound, while knobs and faders provide the possibility to change these properties. Interpreting and modifying this information is the role of the mixer. In a strictly physical sense then, the sound that reaches the audience is not the same as the sound made by the musicians: it is louder, different in timbre and dynamics, etc. The mixer's intervention aims to produce a sound that is nonetheless recognized as the music performed by artists. The relational aspect of the technician's work is central to achieving this goal.

On the non-relational side, sound technicians to perform the music's transformation in three steps. They capture*, 'retouch' and potentially restitute* music.

Music is captured* by microphones, or DI (direct input) boxes in the case of an electric source (see Figure 1.1).

Figure 1.1   Example of a DI-box - photo by Bene

Subsequently sound technicians use different tools, such as a mixing desk, to modify the raw captured* music. This step is called retouch. After being modified, music is restituted* on speakers. Two restitutions* happen at the same time. For the audience, the sound is restituted* at a higher volume than that achievable by musicians alone: this is the front of house sound. Musicians on stage also need a sound source of their own, the monitor* or stage sound. The whole path of the sound, from the musicians' instruments to their ears and those of the audience, is called the audio chain. This audio chain is the process through which material properties of music are transformed to take their final shape, which will be received by the audience (front of house) and musicians (monitor*). This final shape is called an audio image.

Far from being flat and neutral, this transformation cannot be thought of as a direct and simple raising of acoustic pressures, for several reasons. The first one has to do with the physical properties of sound. Indeed, human perception of acoustic pressure does not follow a proportional scale, but a logarithmic one. This means that in certain value ranges, slight changes on the mixing desk have a big impact, while in other ranges, the impact is limited even in the case of big changes. Thus, in order to obtain an audio image that resembles a proportional rise in the volume of the music, technicians have to apply a range of filters and tools, such as gain, which amplifies the signal of an audio source, or EQ (equalization), which boosts or cuts frequencies within the audio spectrum.

Furthermore, the technicians' tools come with limitations. For instance, if a microphone and a speaker are too close, the speaker's sound can be received by the microphone, reamplified, re-received again, creating what is commonly known as "feedback" or the "Larsen effect" (Mercier (dir.), 2017). As the loop goes on, the sound level rises progressively up to the maximum of the system's capabilities, and can cause its destruction. The sound is also generally unpleasant to the human ear, and technicians generally avoid it.

The physical properties of sound and signal treatment therefore require, first and foremost, a non-neutral transformation. The various tools which are combined to form a PA system (microphones, mixing desk, effects modules, cables, amplifiers, speakers…), allow technicians to handle amplification while accounting for the specific kind of changes that these physical properties imply. However, the ways these tools are used are in fact hardly driven by the material properties of the music and sound. Rather, it is driven by the type of relationship that emerges between musicians and technicians, as we will see now by looking more closely at how the front of house sound is constructed.

This fieldwork showed different degrees of collaboration between musicians and technicians for a live show. However, these degrees can be summarized as the range between two archetypes: one extreme case is the purely one-time collaboration, when a technician is hired by a venue to set up a PA system and do the sound of a band he never met before. The other extreme is the long-term collaboration-based pre-existing work relationships, in which the technician knows the artists well and is deeply involved in artistic matters. For the sake of contrast, these two archetypes will be the focus of this chapter's remainder.

Long-Term Collaboration Based on Pre-existing Work Relationships

In all cases, the audio image entails a full reconstruction of the music. Perrenoud (2007) described how recording was an important step in the formation of young musicians, as the shock caused by the contrast between what is heard in rehearsal and what is heard in recording constitutes a challenge to the musicians' identity. Sound engineers, in all cases, have to handle this reconstruction.

In the case of a long-term collaboration, sound engineers work as an external ear guiding musicians into the construction of their audio image. Their role can then be close to the one of artistic director:

> *"The first concert I did with them I said 'wow, catastrophe' (…) it was full of sound problems, arrangements…stuff that did not work (…). Maybe not all sound engineers are like me but I really use the way I feel music. And if I am not feeling well, it means it does not work. So, at the beginning I am a bit lost and I said okay…we have to do something about this. So there, we start a discussion and to make the problem evolve (…)." (FG, interview)*

This quote illustrates how the technician doesn't only draw on their technical expertise, but also on their personal sense of aesthetics to provide advice and to properly use the tools for constructing the audio image. This collaboration was part of the process of delivering an artistic project, that was both artistic and technical:

> *"Z: In fact, you were giving them artistic feedback.*
>
> *Yeah. But not only. Sometimes it is purely technical but also artistic. They know it and they have really been receptive to it. So, on the road, step by step, gig after gig, we will…they will modify stuff…they were looking for their sound. (…) what is interesting for me is to participate in the evolution of this band, you see." (FG, Interview)*

This collaboration can be central to the construction of a band's musical identity. For instance, one of the observed bands played acoustic instruments, but with the intention of making them sound like electronic ones. The role of the sound engineer, in this case, is crucial, the illusion is created through the transformation s/he sets in motion. In this case, the use of technological tools is driven by specific aesthetic conventions. For instance, during the observation, two microphones were pointed at some instruments. In fact, these two microphones had different sensitivities, and were used according to the playing style of the musician, in order to make the instrument sound "electronic":

> *"The drummer plays with all kinds of sticks on the snare, in all kinds of ways… on some parts of the set, I know that he will play loud and so I use the dynamic microphone. On other tunes, I know he will be soft so I use the static." (FN, small talk on the spot)*

This kind of precise setting requires a lot of time and effort. In this particular case, the project required several historical conditions. The technician and musicians met at school around ten years before the performance and had been working together since. This allowed them to develop a specific aesthetic in which the technician was completely involved. As a symbol of that, he was linked to the musicians during the representation by wearing a stage costume, although he was working from the room. He became completely integrated into the band, to the extent that his absence would make it impossible for the band to play:

> *"For instance, with [this band], I've created the thing that if I am not here, they can't do a concert. [...] I think this is what I liked in this job in fact. Being part of the artistic project. And the consequence of that is precisely to be irreplaceable."* (FN, Interview)

This integration goes far beyond the artistic side of the work, and implies his participation in the negotiation of the band's gigs:

> *"When I negotiate with the venue's technician, I need to defend the project: 'Yes, we need a piano, no, we are not putting a keyboard because the guy scratches the strings so a digital keyboard is not gonna work.' To the tour manager: 'yes we really need to give musicians' what they say they need, yes we really need to bring our sound engineer because it's essential'. This kind of things. Yes, we really need two hours of soundcheck... To the musicians: 'yes, we really need to be focused on stage because we really play, we don't play with machines...so no we are not going to make show, we will not run in the audience.'"* (FN, interview)

The ability for him to negotiate in this way is reinforced by the support of other people involved in the project, whether they are musicians or cultural intermediaries:

> *"We are all committed to this project's originality. And that's really cool, that you don't find yourself alone, defending the desires of artists that aren't conscious that they are on the margins, with the managers of your band who fall in with the programmers and who says 'but wait, everyone tells us this is nonsense, why are we asking this?'"* (FN, interview)

If the particular musical identity of this band is partially formed by technologies, the work on sound that makes this identity emerge is mostly driven by the type of relationship constructed between the insiders of the project. The musicians and technician, in this case, knew each other for several years. But how accurate is the reconstruction of the audio image going to be when technicians and musicians don't know each other in advance? Does the absence of a solid relationship allow the material to take over and "impose" its definition of the situation?

One-Time Collaboration With Strangers

When they don't know each other beforehand, musicians and technicians must establish a working relationship during the so-called "sound-check", the moment before the concert when musicians get on stage and play for the sound technician to balance the sound. In this configuration, the communicative and relational aspects of the soundcheck are particularly apparent, as working routines are not settled between musicians and technicians. If the musicians do not give information spontaneously, technicians have to ask for all relevant information. For instance, during an observation of a three-day open-air festival*, a cumbia big band brought nine persons on stage and did not bring a technician with them. The leader of the band, clearly aware of this situation, brought a set list to the technician half an hour before the beginning of the show and explained to him who will play solos and when plus how percussion and melodic instruments, as well as lead and choir voices should be balanced. In this case, the sound design was the result of a direct negotiation between the musician and the technician, under the mantra of "the show must go on": "give me what you have and I will do what I can".

In contrast, during the same observation, one of the bands came without a technician and did not communicate with the front sound engineer at all. They were playing as if they were rehearsing and stopped the soundcheck suddenly and without warning. They then went off stage, leaving the front of house and monitor* sound engineers with a sound that had to be completely reshaped during the first minutes of the show. As a consequence, the sound quality was initially poor and only gradually improved during the performance itself.

When they do not know the musicians, technicians have to grasp the musical identity of the band within a few minutes on the basis of the tunes that are played in the soundcheck. Except for trivial and very conventional moves such as putting reverb on a voice, technicians usually do not allow themselves to aesthetically change the music. To produce the sound of the band, they will rely on an interpretation of how the band is supposed to sound, according to what can be expected from how the musicians look on stage, based on the information they already have and using their own experience as technicians, and of course based on what the musicians tell them. In this situation, communication and 'reading' other persons are the keys to producing the right sound.

Multiple nuances exist between the two extremes of long-term collaboration based on informal pre-existing work relationships and unique one-time collaboration. Each variation leads to differences in the material form in which music will be perceived by audiences. One recurring example was how a label, a manager, or another cultural intermediary would assign a technician of their own, who did not know the musicians before but potentially engaged in a more or less long-term collaboration.

In sum, the construction of the audio image of musicians on stage largely depends on the type of relationship they have with technicians, and on the level of communication between them. Physical properties of sound and gear characteristics are minimally influential in performing this task. However, getting behind the mixing desk and mixing is only a part of technical work in live music. Before that, among other things, the PA system needs to be selected, set up, and tuned to the acoustic environment in which it will play. Due to the object-related nature of this task, it can reasonably be expected that this part of the job will be more driven by objects, environments, and tools.

Non-relational Technical Mediation: Producing Homogeneous Sound Throughout the Room

Audio images may be conceived in rehearsals or in studios, far from the actual place in which they will be played. Thus, besides balancing the different music sources during the soundcheck, technical intermediaries of a live performance have to ensure that the audio image can be properly reproduced in the physical environment of the concert. For this, the choice of the PA system, its position in the venue, and its adjustment are crucial elements that happen under the responsibility of technical intermediaries. The question asked here is to what extent these choices are guided by the physical properties of material environments, or by the interpersonal characteristics of art worlds.

First, it has to be noted that the mixing desk is a tool that can compensate, to a certain extent, for changes in the environment. For instance, in the case of a band on tour, if today's room resonates* at 2.5 kHz and that was not the case in yesterday's room, the frequency can be reduced with the help of the mixing desk. But mixing may not be enough to address the challenges posed by an acoustic environment. Indeed, acoustic environments can distort the work of the sound engineer and harm the production of the desired audio image:

> *"The problem is that we were playing this project in places that were not at all suitable. That was a very rock project, very loud. Ultra rock'n'roll. And we were playing it in like, theater venues...we did it in [a classical music venue]. There's four seconds of reverb. [...] I said 'I won't do the front of house. It will be horse butchery, that's out of question'. I said: 'If you want me to do the monitor*, I'll do the monitor*, that's*

cool.' But doing rock'n'roll in a venue with four seconds of reverb, it's suicide. It will not sound good in front, that'll be hellish, and I don't want to do that." (FP, interview)

In this example, there was a gap between the ideas proposed by the producers and the material constraints of their project's implementation. This gap can be reduced by appropriate use of the PA system technology, but according to the respondent, it was not possible in this case. The room's acoustics would have prevented him from getting a satisfying result.

The general purpose of the sound reproducing device is to produce a homogeneous sound for the whole audience, to bring the same audio image of the music to everyone in the room. This is an ideal, and impossible to reach in practice, a typical example where the "point in which a production system meets the vagaries of the material world" (Barley, 1996). The design of the PA system itself is a way to get closer to this ideal. In the 1990's, the technology of Line Array was introduced, leading to a substantial change in concert amplification. Rather than concentrating the sound diffusion in one point that targets the whole audience, the Line Array multiplies the sources so that each is focused on one part of the audience only (Figure 1.2).

Less reverberated, and thus less distorted, sound reaches the audience's ears, allowing them to perceive much clearer sound coming directly out of the speakers, and which is under control of the front of house engineer:

"Our goal as system engineers, is to install speakers that will spray...well that will be directed to the audience, where there is audience. The purpose is to really only focus where there's an audience to avoid sound dispersion, that will create problems...

– Yeah, if you spray the walls...

– Exactly, you've got reverberation. It blurs the direct sound and all." (FQ, interview)

This task of tuning the PA system is a constant in all observations. The system may need to be set up, for instance in festivals* or other such temporary events. It is permanently installed in venues that stage concerts regularly. However, the system is always "calibrated"*, meaning that the front of house engineer ensures that there is no distortion caused by the system itself. In the observations used in this chapter, it was generally done by playing a piece of music that the engineer knew well, comparing what was heard from what was expected for the tune to sound good. But technology can also be used, especially in big productions. A "sine sweep" has been used in two observations. A measure microphone is placed in the room, and a sinusoidal sound is played at all ranges of frequencies. The

Figure 1.2 Example of a
Line Array system - picture
by Rudolf Schuba

graph that results (Figure 1.3) is analysed and the system calibration is modified
in response to what is displayed.

However, the sound engineer remains the final decision-taker:

> *"The final judge is always your ears... You can have all the analyzers you want, I use
> my ears to finish calibrating a system...Everything may look good on the analyzer, and
> then you have complete shit coming out. [...] 2, 3% of the time, you have systems that
> react...that's really weird, you have systems that don't react as usual. They will give
> you a flat response on the analyzer, you have a sweet response in flat frequency, but
> with your ears it sounds like some systems give you more high frequencies, despite the
> empirical measure [..]. There's something like, a third factor, not on our analyzers. I
> don't know exactly what this is, but it's always worth listening to ensure that it's not
> ruining everything." (CA, Interview)*

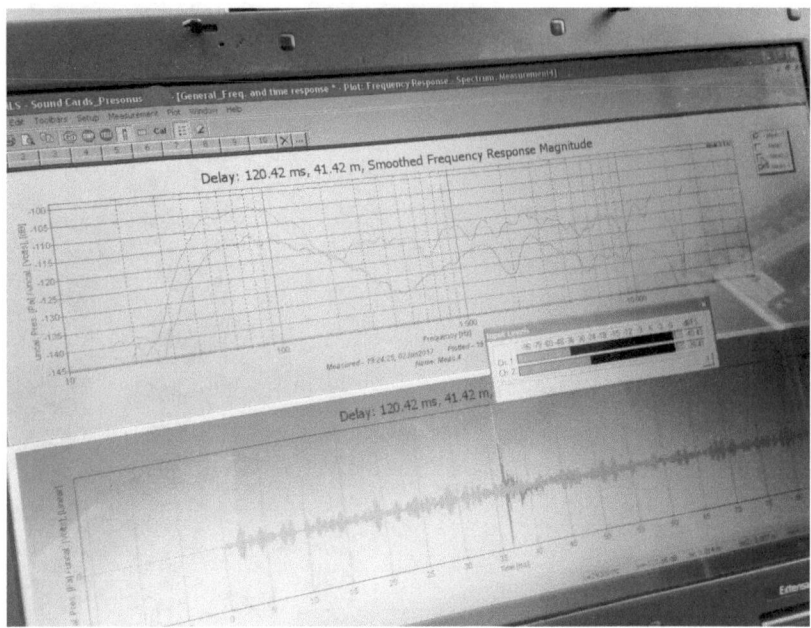

Figure 1.3 Graph resulting from the test of a system through a sine sweep - picture by the author

The person quoted here works at the highest level of international concerts, where the most advanced technology available worldwide is used. In this case, technology works to assist the subjective feeling of a trained and experimented professional. Hence, we can see that the influence of education and experience of technical intermediaries prevails over technological choices in practical processes, whatever the range of technologies used. However, before relying on such technologies, one has to need it, and to be able to fund it. In big shows, handling both the tasks of mixing, and setting up the PA system is too complex for a single human being. In this case a specialist in setting up can be appointed, if this is financially feasible:

> *"I am dealing with the diffusion system for the audience. [...]* **That's for big tours.** *You have a technician, a system engineer who handles the diffusion system." (FQ, interview)*

Hence, proper tuning of PA systems to material conditions relies on two factors. First, it relies on the means available to producers, that give the possibility to acquire technologies and the people able to operate them. Second, it relies on educational systems training technical intermediaries, whether through education or professional experience. These systems develop the abilities to properly employ technologies, both in terms of using them to their full potential and of the ability to be creative in their use when faced with concrete problems. Moreover, they also inform technicians of the limitations of the gear they are using. The importance of this theoretical, practical and empirical knowledge is illustrated by this obser-vation of a festival* organized in a church. The programming included a good share of amplified pop and electro music. The ten seconds reverb of the church was impossible to ignore. Hence, in order to provide a clear sound, the sound engineer, a person in his 30s who benefited from one of the most advanced edu-cation programs in sound engineering in France, had to place delayed speakers in a precise pattern:

> *"In this place, the problem is that the reverberated sound will quickly overcome the direct sound you hear from the speaker. 10 meters from it and you're done, basically. You hear more reverberation than the sound of the speaker and you don't get a thing. So the idea is to put a recall (rappel) speaker at 10 meters from the stage, and another one 5 meters further, and another one, etc... in this way everyone is always covered by a speaker."* (Sound technician, small talk on the spot during observation)

He explicitly said that he was not "fighting" the reverberation of the place, as it would have resulted in a "dull" sound. He integrated it to a degree into his mix, as a part that he was not able to directly control with a fader, but indirectly through other methods. This subtle use of technology in order to manage an acoustic environment rather hostile to a pop concert was facilitated by his training, and only required casual gear to be implemented. Another observation, happening in a semi-cylindrical hall 20 meters high, had the same reverberation problem and the same kind of speakers, but none of the engineers came up with the solution of spreading speakers across the room. As a result, the mixer did not manage to produce the intended sound.

The work of technical intermediaries appears to be the answer to the con-tradictions inherent in the materialization of a project's design. Problems appear when the environment constrains the producers' intentions. While material pro-perties of the environment are indeed considered during calibration, the ability to cope with them depends on the material and human means available to materially implement the project. Economically, money is needed to pay for workers and materials. Practically, workers must be able to fully exploit the potential of technologies. This ability depends on the quality of their training, whether through primary education or professional transmission, coupled with an accurate distribution of responsibility. The picture, then, appears to be much more complex than a dialogue between humans and materials, in which the latter heavily determine how the situation issues.

Material Constraints and Conventions: Technical Ability as the Locus of Agency in the Interaction Between People and Things

Material and relational influences on technical work have been thus disentangled through the study of the working practices of technical intermediaries on two central tasks leading to a concert production. This analysis shows that the work of technical intermediaries is mostly driven by relational factors during a live music performance. The material environment provides a response to the intentions of people in the art world.

In an art world, conventions are the cornerstones of a successful performance: they allow musicians, technicians, cultural intermediaries and audiences to find common meaning in the event they are participating in: concerts in the cases studied here. Cultural intermediaries produce a symbolic frame for the musicians' work, thus relaying it to an audience in a form that makes both economic and symbolic sense. They use cultural references and social skill to mobilize distribution channels in order to organize a successful connection between artists and audiences (Becker, 1982; Lizé, 2016; Maguire & Matthews, 2012; Negus, 2002). This organization is in itself a production system, in the way described by Barley (1996). Conventions of this cultural system constitute a middle point between the artist's aesthetic views, the cultural intermediaries' needs and the audience's taste. But conventions are nearly independent from material properties of the environment in which a performance takes place. Instead, conventions are constructed through a process happening before the concert. Although it might be at some point rely on material elements, this process goes far beyond the material conditions in which the concert actually happens. Commercial practices (Becker, 1982, p. 46), power relationships (Becker, 1982, p. 47), and cultural practices (Becker, 1982, p. 48), are some examples of the larger processes Becker mentions as crucial for explaining the adoption of a convention by an art world.

Material constraints, on the other hand, are always attached to a particular time and place. Each room sounds different, each PA system has its own characteristics, each microphone transforms the sound in its own way. Moreover, the same place will not sound the same at two different moments in time, as changes in the weather and particularly the humidity level, or the presence and size of the audience will transform the way sound propagates through air.

The role of technical intermediaries is precisely to shape the material properties of both the environment and the music in order for conventions to be applied independently of the environment. The technicians studied here have an objective: to rebuild an audio image of the band on stage, defined by the consensus reached beforehand by other actors involved in the art world, which is shaped by and expressed through artistic conventions. In order to achieve their objective, technicians must account for the material properties of the environment in which this image is built, and they rely on specific tools that are only used by themselves in their working environment. In other words, the expectations of the shape of the audio image are defined independently of the environment in which the concert takes place, and technicians will have to deal with them in order to provide an image that fits these expectations. As they are using specific tools for this purpose, they indeed form a kind of association between humans and non-human entities, as formalised in Actor Network Theory (Callon, 1986).

But this chapter's observations call into question the notion of generalized symmetry at the core of this theory. Indeed, it is clear that the function of sound engineers is precisely to neutralise, as much as possible, the effects of material variability, so that essential conventionally-meaningful aspects of the audio image are not modified. Environments are a given parameter to them, around which they will have to work to achieve a purpose defined independently of these environments. For that, they will implement a strategy based on their education and experience, for instance through the integration or suppression of reverberated sound in the audio image. Clearly, therefore, objects and environments do not take decisions by themselves, and will not purposefully implement a strategy to counter the sound engineers' plan. In this sense, environments, tools, and people cannot be considered as equal participants in a network. The "negotiation" of non-human actants results in this case from the framing work of technical intermediaries that integrate the material in the network. Assuming generalized symmetry erases this work from the analysis, and thus overestimates the influence of objects.

This is why the notion of *technical ability* appears more accurate than material agency. Technical ability (in this case the ability of technical intermediaries to materialize cultural objects according to conventions) is the human ability to address and manage objects' and environments' physical reactions. In other words, technical ability is what determines the human response to the reactions that materials will produce in response to human intentions. Technical ability defines the capacity to successfully adapt the means to the challenge posed by a specific context. In this fieldwork, the ability to do so relies on technical intermediaries' capacity to understand the intentions of other human actors of the art world through communication, and on their knowledge and experience of the tools they are using to materially achieve these intentions. Both capacities are constructed by socialization, a concept that relies both on structural and interactional theoretical perspectives.

To use a mathematical metaphor, human agency on objects and environments is thus the result of an equation in which the variables are technical, imaginative, and economic resources and abilities. In this equation, the properties of the material environment must of course be included, but as a given. In social analysis, the material world typically is a constraint rather than a determining variable. Accounting for physical properties of the environment in the same way as interpersonal interactions in social sciences, is problematic theoretically (Elder-Vass, 2008, 2015) and practically. It can lead us to overlook human agency over material constraints, built through the social organizations or institutions invented to face collective challenges. Therefore, it can lead to an overestimation of the material's impact, and worse, an underestimation of human ability to face material challenges. Finally, it can make us ignore that behind any action intended on the material there are intentions and purposes that are socially constructed, and effectively independent of any physical law. Technology, if it is indeed associated with humans and materials, is essentially a tool that serves these purposes and that needs to be adapted to it. In a professional context, such as a concert production, this task is the one of technical intermediaries. Therefore, "material agency" depends more on their work and the mediation that they produce rather than on the material physical properties.

Conclusion: Music as Meaning, and Technology

This chapter analysed the working practices of live music sounds engineers, a particular category of technical intermediaries. By questioning the notion of "support personnel" (Becker, 1982), it presented a model of art worlds as constituted of four kinds of actors: artists, audiences, technical and cultural intermediaries. It showed the specific role of technical intermediaries, which broadly consists of modifying the physical shape of cultural objects in order to make them correspond to the conventional expectations of their art world. Furthermore, its analysis

questioned "generalized symmetry" postulated by ANT (Callon, 1986), and showed that "the agency of objects" was more accurately captured by the notion
of "technical ability", i.e. the capacity to mobilize social, economic and cultural
resources to face material challenges. It identified the economic and social means
of actors of the art world, the education and experience level of technical intermediaries, the types of relation and quality of communication between artists and
technicians as primarily influencing technical ability, against physical properties
of objects and environment, or available technologies.

These results have a number of wider implications for our understanding of
cultural performances, and the agency and social role of objects.

First, Becker's notion of "support personnel" does not adequately capture technical work in art worlds. Indeed, more than simply delegating tasks that artists or
other members of the art world could do themselves but do not have the time or
interest in, technical intermediaries provide crucial input during a creative process,
as well as during its performance. They bring a specific technological competence
that is necessary to perform the cultural activity, and they use this competence to
make autonomous and potentially creative choices which impact the final shape
of the cultural object.

Technical intermediaries are present everywhere where artistic content must
be modified in order to fit the frame designed by artists and cultural intermediaries for a presentation to an audience. The typology of art worlds proposed
here does not intend to lock actors in specific and rigid roles. It rather aspires
to describe the different actions necessary to produce a cultural object or performance charged with meaning. Hence, it provides an appropriate framework to
study comparatively the distribution of tasks, power and responsibilities in different art worlds. In this framework, technical intermediaries provide a good entry
point to understand other actors' expectations. Positioned at the last step of before
the presentation of the performance in front of an audience, their work combines
all the attempts from different actors to shape the cultural object in the way they
think it should be according to their own position, making them easily observable by the researcher. This position, however, can potentially be found in other
fields than the cultural one, fields where the process of transforming designs into
material objects is also found.

Second, the critique of generalized symmetry has led to coin the new notion
of "technical ability". This concept allows to avoid anthropocentrism in social
sciences, allowing us to understand the agency and importance of objects without
falling into the trap of material determinism. It opens a perspective that accounts
for the influence of objects and environments but focuses on how society organi-
zes itself in order to integrate them. Thus, the concept calls for further exploration:

how is technical ability improved or diminished by social processes? How do previous interactions influence technical ability in a defined moment? How do power relationships influence technical ability? Such questions give us a new view on human–material interactions to provide different perspectives in pragmatic and critical sociology. Some of these questions will be addressed in the following chapters.

A Comparative Approach to Making a Living From Sound Engineering

2

Professional sound engineers need to make a living. Like many artists, they are generally employed on short contracts, and they tend to work with several employers (Menger, 1999). As a result, they cannot rely on a regular income. Jobs come and go, and they have to find a new position every time a short-term contract ends. Moreover, as independents, their wage is defined by their value in the labour market, and not by a regulation defining, for instance, a minimum wage for the profession. In other words, their labour market is commodified (Greer, 2016) and they do not benefit from a stable employment relationship (Bosch, 2004) to ensure their material living conditions.

However, in this regard, the French and Dutch institutional contexts are very different. Indeed, in the Netherlands, the labour market of freelancers in performing arts is highly commodified: technicians work as independent contractors and can be hired under 0-hour contracts. Conversely, the French equivalent labour market is marked by the presence of a specific regulation, which brings greater income security through the presence of unemployment insurance which compensates for periods without work, and guarantees coverage by public health insurance.

While this regime's structural deficits led scholars to question its sustainability (Menger, 2015), I would argue that it indirectly induces a transaction which makes the sectors responsible for the performing arts "cost disease" (Baumol & Bowen, 1966) compensate its effects through a form of "polluter-pays" logic. In France, through the unemployment insurance of freelancers in performing arts (*Intermittents du spectacle*), sectors with high productivity gains fund the financing of the

---

Electronic supplementary material The online version of this chapter (https://doi.org/10.1007/978-3-658-33029-3_2) contains supplementary material, which is available to authorized users.

A. Battentier, *A Sociology of Sound Technicians*, Musik und Gesellschaft, https://doi.org/10.1007/978-3-658-33029-3_2

benefits of a stable employment relationship for freelancers in performing arts. Baumol and Bowen argued that their law could only be broken by subsidies, whether public or private, in order to maintain the economic viability of performing arts activities. In this chapter, I will argue that the transaction mediated by the French unemployment insurance for freelancers in performing arts constitutes another way to break Baumol's law.

Sound technicians are required in order to be able to stage musical performances. It is therefore necessary to anyone who wants to do so to have access to the economic means to hire them. While hiring them as freelancers constitutes a way to lower the cost of a performance, I will also argue that the re-commodification of the labour markets of artists and technicians implied by the French system is a way to facilitate the material possibility of the interaction rituals that are musical performances.

Sound Technicians and the Baumol Law

Published in 1966, *Performing Arts: the economic dilemma* (Baumol & Bowen, 1966) established a relationship between the outstanding increase in productivity in the US private sector, mostly composed of industrial activities at the time, and the structural economic deficit within performing arts activities. As technology allowed the production of more items from the same amount of work, benefits to industrial companies increased, followed by an increase in the wages of workers of the sector. As salaried consumers became able *en masse* to spend more money on goods and services, the general cost of living increased, which created the necessity for employers across the country to raise the wages of their workers, whether their activity benefited from productivity gains or not. Performing arts, in this matter, are a specific activity: they rely on a resource that cannot achieve significative productivity gains. Indeed, it relies mostly on the skills of its personnel, whose outputs cannot be improved by technological innovations. Contrary to industry, due to the nature of the activity, the amount of human work necessary to produce a performance cannot be reduced by innovative mechanical processes.

A tension thus appears between the necessity to raise the wages of performing arts' personnel, and the impossibility of covering the subsequent increase in personnel costs by a corresponding rise in income. The authors show the presence, in US major orchestras, opera companies and Broadway theatres, of an "income gap", a structural deficit that theoretically condemns performing arts to the impossibility of economic self-sustainability. This income gap is interpreted as the symptom of the "cost disease" of performing arts, which can be cured by providing money without expectation of direct financial return, whether by the public sector in the form of subsidies, or by the private sector in the form of philanthropy. The existence of an income gap has been confirmed by more recent works on US (Felton, 1994; Flanagan, 2012) and Canadian orchestras (McGrath, Legoux, & Sénécal, 2017). The cost disease, however, appears not be limited to the performing arts, but extends to the more general services sector, where productivity tends to rise more slowly than in manufacturing industries. It has indeed been detected in US education and health care sectors (Bates & Santerre, 2013), as well as Norway's defence and public administration sectors (Borge, Hove, Lillekvelland, & Tovmo, 2017).

Baumol and Bowen have identified subsidies or private donation as the only possible way to fill the income gap and maintain performing arts activities. However, employers can attempt to address ever-growing personnel costs by externalizing them. Performing arts are known to function on project-based employment: artists, technicians, and potentially cultural intermediaries are hired on short-term rather than permanent contracts (Hesmondhalgh & Baker, 2010; Menger, 1999; Throsby & Zednik, 2011). Hence, during a decrease in work caused by an economic decline, employers just stop hiring freelancers, and do not have to support the cost of wages. Economic risks are thus assumed by employees, who are deprived of income until activity resumes (Menger, 2015). Such characteristics of performing arts, and artistic labour markets in general, is considered to have been the impetus for a recent development in capitalist economies (Boltanski & Chiapello, 2011), that has led to a degradation of the standard employment relationship and precarization of the general labour market (Bosch, 2004).

Nevertheless, French law counters the precarization of freelancers in performing arts with a specific hiring and unemployment regime, which has no equivalent elsewhere in the world (Menger, 2015). The law states that permanent contracting is the general norm of hiring, and that non-permanent contracting is an exception that needs to be justified. Hiring freelancers on a regular basis in the performing arts sector is justified by the "specific nature of the activities of the sector"[1]. While the law acknowledges a specific need for flexibility from employers of the sector, it also provides social security to freelancers. First, a working contract is established, which formalises the responsibility for the employer and the employee to contribute to public health, retirement, unemployment insurances, thus granting the latter a right to welfare benefits. This contrasts with the regular situation of freelancers, who are not contractually employed but are legally considered service providers for a client, assuming their own health, unemployment,

---

[1] Article D1242-1 of Labour Code

and retirement insurance. Second, French freelancers in the performing arts fall under a specific regime of unemployment insurance. While the regular regime considers that a worker is either employed and thus salaried, or unemployed and thus can claim unemployment compensation, freelancers in performing arts can do both at the same time. Once they have accumulated 507 worked hours in 12 months, they will receive financial compensation for the days on which they do not work for the next 12 months. In sum, they have the right to unemployment benefit despite changes in workload, and they can accumulate it with their actual salary. Hence, once the threshold of 507 worked hours is passed, they secure an income for the year to come, granting them a fundamental feature of a standard employment relationship.

In contrast, in the Netherlands, freelancers in performing arts are considered like any other freelancers. They fall under the regime of "independent without personnel" (*Zelfstandige Zonder Personeel*): they are micro-entrepreneurs who do not receive a wage from an employer but a payment remitted by a client after they submit an invoice for a service. In this regime, they must provide for their own social security. They can have a slightly better status if they are hired on a "0-hour contract": they are then employees, but their employers have the right to decide whether they have any working hours at all. As a result, even though they have retirement contributions and holiday leave, 0-hour employees do not have any guarantee of income (Vonk & Jansen, 2017).

From the point of view of employers, this externalization is a way to tackle personnel costs, which is arguably made necessary by the effects of Baumol's law. However, its effects have never been accounted for in the field of popular music. Cowen (1996) even argued that popular music constitutes a counter-example of the archaic nature of performing arts sectors assumed by Baumol and Bowen. He argued that performing arts, and particularly today's "cultural winners, such as rock and roll, country music, and heavy metal" (Cowen, 1996, p. 211), can benefit from innovations in processes and in product. The former is allowed by technologies like recording and sound amplification, which make a performance to reach more individuals than before. The latter comes from the creative nature of performing arts, which can attract new and numerous audiences by diverse artistic innovations. Furthermore, other studies focused on alternative economic models trying to tackle these effects: two documented examples are the development of festivals* (Frey, 1994), which both reduces production costs and concentrates performances at a moment of higher audience receptivity (generally holidays), and hybridization of economic models (Hitters & Richards, 2002) that allow economies of scope, by concentrating diverse artistic activities likely to attract more audiences, or by coupling a lucrative activity (such as bar or restaurant) with performances.

Cowen points out an "unjustified bias [of cost-disease proponents] towards high culture", considering that Baumol and Bowen have not considered all forms of performing arts, and particularly popular music (Cowen, 1996: 211). The data gathered for this dissertation, through the respondents' narratives and the financial figures of pop music venues in France and the Netherlands, brings an opportunity to study Baumol's law in popular culture. It will allow us to ask to what extent popular music is affected by Baumol's law, thus showing its generalizability beyond the realm of highbrow culture. It will also give us information on the role of personnel externalization into tackling these effects, and on the effects of two opposite institutional ways to deal with the social security of externalized personnel. Finally, it will show the general economic context in which sound technicians are embedded in order to produce musical performances, thus paving the way to understand how their concrete working practices and relationships are framed.

Comparing the Structure of Two Live Pop Music Fields

In order to answer these questions, I will analyse the financial structure of pop music venues in France and in the Netherlands. Specifically, I will compare the extent to which music venues, on a national scale, rely on autonomous income, subsidies, and work externalization.

For this, I will analyse the financial figures issued from the annual surveys performed by the *Fédération des Lieux de Musiques Actuelles* (FEDELIMA, 2016) and the *Vereniging Nederlandse Poppodia en Festivals* (VNPF, 2016). Both organizations are associations of live pop music venues, aiming to share information and coordinate actions in order to develop their activities. They perform a yearly informative survey of their members, in order to establish statistics in the field of live pop music. These two associations coordinate their surveys in order to build comparisons at a European level, through a larger network called *Live DMA*, which unites similar organisations in 15 countries. I have been granted access to the detailed figures of activities, workforce, and finances of 42 venues members of VNPF and 112 FEDELIMA members for the period 1999–2016.

Ideally, a longitudinal study would account for the evolution of incomes and overheads. However, the relatively recent implementation of the surveys implies that many cases are missing, that questions have changed substantially over time, and that analytical categories' boundaries have been blurred from one year to another. Harmonization between national surveys has been realized only for the last few years. Despite trying to apply various inference methods for missing cases, and attempting to create analytical categories that would cover the same concept over a sufficiently long period, the data did not permit a robust comparative longitudinal study. Hence, the strategy adopted aims to compare the financial figures of all respondents in the year for which questionnaires were best refined and harmonized, which is 2016.

I will study the finances of French and Dutch popular music venues in 2016, after decades of increasing personnel costs. Indeed, the productivity gains of western Europe has followed a growth trend similar to that of the US in the 20th century, although European countries' productivity was always a step behind the US and trying to catch up, and also experienced a lull in the 1980s (Broadberry & Crafts, 2009). The last available French and Dutch GDPs[2] (Figure 2.1 and

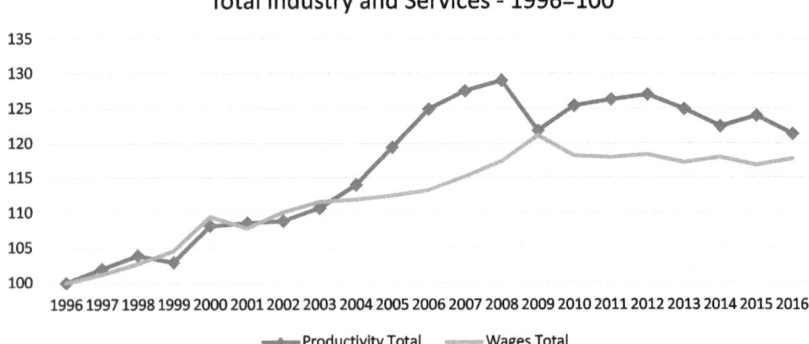

Figure 2.1 Evolution of productivity and wages in the Netherlands, 1996–2016, 1996 = 100. *Source* Centraal Bureau voor Statistiek

---

[2]To compound these graphs, I used national GDP figures, amount of hours worked and wages by branch, converted into purchasing power parity, available from the websites of *Institut National de la Statistique et des Etudes Economiques* and the *Centraal Bureau voor de Statistiek*. Longitudinal figures are available up to 1990 in France and 1996 in the Netherlands. Baumol and Bowen based their study on the productivity of industry because it was the

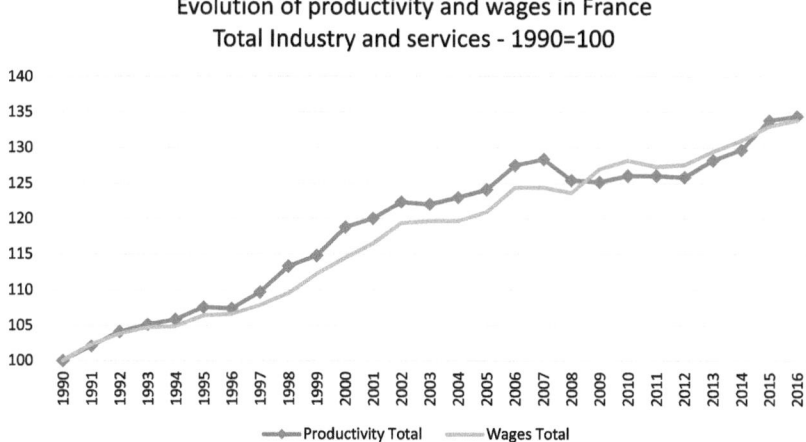

Figure 2.2 Evolution of productivity and wages in France, 1990–2016, 1990 = 100. *Source* Institut National de la Statistique et des Etudes Economiques

Figure 2.2) indicate that French productivity gained 34 index points between 1990 and 2016, and Dutch productivity gained 28 index points between 1996 and 2008 and stagnated since the financial crisis after that. Therefore, the dynamic leading to the cost disease is well present in both France and the Netherlands. The current state of pop music fields must be considered as being both the result of decades of accumulated high levels of pressure on personnel costs induced by productivity gains plus the current lower pressure. I will now proceed to the analysis of how Dutch and French pop music venues currently deal with the rising costs, by relying on subsidies, autonomous income, and work externalization.

---

main component of the US' economy at the time. However, contemporary French and Dutch economies have switched towards service economies, in which the share of industrial activities in GDP, although still important, has drastically reduced. This is why I chose to analyze the productivity of industry and services, measuring productivity and wages of the branches B to E (Industry), and G to N (Services) of the standard industrial classification of economic activities of 2008. In total, these sectors represent 66% of France's 2016 GDP, and 63% of the Netherlands' GDP. A significative rise in wages of this sector is thus assumed to induce the rise in personnel costs described by Baumol and Bowen. However, for both countries, the productivity of industry and services measured independently follow notably similar trends in the period studied.

Live Pop Music Venues' Economic Models

Subsidies

Table 2.1 Subsidies for pop music venues in France and the Netherlands. *Source* VNPF (Annual Survey 2016) & FEDELIMA (Observation Participative et Partagée 2016)

	France	The Netherlands
Share of public subsidies in total income	59%	28%
Share of the state in total subsidies	16%	2%
Share of intermediate levels (regions, department, provinces) in total subsidies	18%	1%
Share of municipalities in total subsidies	58%	97%

The first result is that, on a national scale, pop music venues rely on extensive public funding to function. This is particularly true of French venues, whose average income is 59% from public subsidy. However, Dutch venues also depend on subsidies, which comprise 28% of their income. Even though recent Dutch policy documents insist on private contributions to fund culture, private contributions to pop music venues are minimal in both countries. Venues can benefit from three kinds of public subsidies: operating, project and employment. Operational subsidies are intended to fund the daily running cost of the venue. Project subsidies are intended to help with a specific investment that the venue must justify, such as an extension to its building. Employment subsidies are to fund part of the workers' wages. In both countries operational subsidies represent the largest share of this type of income. They represent 96% of Dutch subsidies, and 85% of the French ones. For the latter, the rest is mostly covered by project subsidies (11%).

This distribution reflects national cultural policies. In France, these policies were first defined nationally, with the creation of the ministry of culture in 1959 (Perrin, Delvainquière, & Guy, 2017). Its general guidelines were for the democratization of legitimate culture, and thus its scope did not include popular music at this time. These guidelines were changed in 1981 by the minister of culture Jack Lang who implemented policies of financial support and legitimation of popular arts. These policies were coupled to devolution acts* intending to decentralize France's public action, which establishes equal competence at different levels of governance in cultural matters. As a result, music venues are perceived, and perceive themselves, as fulfilling a public service mission, as explicitly specified in

1998's *Charter of the Public Service Missions for Performing Arts*, and formalized in a labelling policy specified in the *Note-Circular of 31st August 2010 (modified) relative to the national titles and network in the performing arts*. This mission goes beyond concert production and includes support of local artists as well as inclusion of new audiences.

The Netherlands followed a similar trend: subsidizing orchestras and opera companies starting from the 1950s, and recognition of popular music as worth public funding starting from the 1980s. However, these policies were influenced by a certain reluctance to directly involve the state in cultural matters, which are generally considered more as a private matter than in France (Hamersveld, 2009; Van Der Leden, 2017). Subsidies for artistic activities are granted by independent committees, but funded with government money, on the basis of "artistic quality", the definition of which is at their discretion. During the last ten years, the amount of money funnelled into cultural policies has dropped, notably with a 20% cut in 2011 (Meerkerk & Hoogen, 2018). Cultural entrepreneurship became a central guideline of these policies. Following the trend of considering culture as an attractive factor, initiated by the works of Richard Florida (2005), municipalities increasingly funded infrastructure like pop music venues in order to attract populations perceived as vectors of growth. In the specific area of popular music, the March 2016 letter from the minister of culture *Een investering in popmuziek* (An investment in pop music) stresses the importance of filling venues, and thus implicitly the importance of acquiring income from box office.

These different visions on the governance level at which cultural policies must be taken explain the differences shown in Table 2.1. While almost all subsidies for pop music venues are granted by municipalities in the Netherlands, funding sources are more distributed in France. Furthermore, publicly funding cultural activities does not trigger significant suspicion in France, where such funding is perceived as a way to escape cultural uniformization resulting from market dynamics (Poirrier, 2002). This higher acceptation, coupled with the political framing of state intervention as a public service, and the higher budget cuts in the Netherlands, explain the greater reliance on subsidies of French venues. This difference, as we will see now, has an impact on the activities hosted by pop music venues in each country.

Autonomous Incomes

Following the guidelines of recent cultural policies, Dutch venues tend to be more specialized in the production of musical performances. Their economic model

Table 2.2 Main sources of autonomous income of pop music venues in France and the Netherlands. *Source* VNPF (Annual Survey 2016) & FEDELIMA (Observation Participative et Partagée 2016)

	France	The Netherlands
Proportion of total income coming from box office	15%	33%
Proportion of total income coming from food and drinks	10%	25%

depends more on income from the box office: while only 15% of French music venues' income comes from concert tickets, it is 33% of Dutch venues' income. Baumol and Bowen showed that ticket prices cannot really be increased to raise the box office income to the point that the cost is covered. This appears to be confirmed by the fact that the difference of the share of the box office income between France and the Netherlands corresponds to a difference in the volume of programmed events. In 2016, 112 French venues programmed 7033 events, an average of 63 events per year and per venue, while 42 Dutch venues programmed 12324 events, an average of 293 events per year and per venue. It is nonetheless worth noting that while the mean number of programmed events by venue is 4.7 times higher in the Netherlands, the mean population density there is 4.2 times the French one, thus offering an outlet for such a dense programming. Besides, in line with Dutch preference for club music (Brandellero & Pfeffer, 2015), Dutch venues notably program a significant quantity of dance events (24% of programmed events), which are less costly to produce than other event types (Hesmondhalgh, 1997). Dutch venues also depend more on income from food and drink sales (25% of total income) than their French equivalents (10%). Furthermore, 21% of Dutch venues have a bar/restaurant opened whether there is a concert happening or not, while this is the case for only 15% of French venues (Table 2.2).

However, French venues propose a range of various activities besides the production of concerts: teaching, socio-cultural actions, artistic residences. These activities tend to not be lucrative, but subsidies are conditional on their presence and they are framed as activities of public interest, supporting local cultural initiatives. As a result, 99% of French music venues pursue at least one of the following music-related activities: support to production, coaching of musicians, sociocultural action, musical teaching, rehearsal rooms, recording studio, audio-visual productions, administrative or technical training, project management, information centre. The difference with the Netherlands is striking, as only 29% of Dutch venues host at least one of these activities. On the programmed activity *per se*, 52% of French venues have programmed events that are not musical, such as theatre, movies, exhibitions, while it is only 21% of Dutch music venue programming.

Autonomous incomes of music venues appear to be aligned with the level of subsidies granted in each country. French venues tend to develop several non-lucrative activities related to music support, sociocultural actions, and artistic diversity, as is required by the mission sheet which comes with public funding. Dutch venues' model of activity is more centred on the staging of musical performances. More than half of their income is autonomous, and their yearly programming is approximately five times denser. This appears in line with the requirements of Dutch cultural policies, especially the ones developed in the last ten years. Hence, we can note that subsidization, as a way to tackle Baumol's law effects, has the consequence of granting a certain degree of control to the funder over venues' activities.

Employment Externalization

Subsidies are thus widespread in French and Dutch music venues. They constitute the answer to Baumol's law identified by its authors, and influence the commercial practices and scope of activities of venues. I will now turn to the other potential answer I identified, which is reducing personnel costs by externalization.

The first form of externalization is in the production of the shows themselves. Indeed, unlike the venues studied by Baumol (main US orchestras, opera companies, and Broadway theatres), the venues I studied do not assume most of production costs themselves, and do not directly employ the artists who perform on their stages. Instead, they tend to buy shows produced by external organisations, such as tour managers, labels, or self-produced artists. They either rent their room and personnel for a fixed price, produce, or co-produce the show. In the first case, the external organisation assumes the risk of an empty room and has to pay the venue anyway. When venues produce shows, they hire artists directly and take on the performance's organization, but this is the exception not the rule. In the former case, which is the most widespread, the risks and benefits are distributed according to a preliminary agreement. All performances in pop music venues are produced according to one of these models, which represent in itself a form of work externalization. Indeed, external organisations assume the costs of the creative processes, whereas venues assume the costs of infrastructures. Buying performances represents 33% of Dutch venues' expenses, and 37% of the French venues' expenses (Table 2.3).

Table 2.3 Share of programming and personnel costs in total expenses. *Source* VNPF (Annual Survey 2016) & FEDELIMA (Observation Participative et Partagée 2016)

	France	Netherlands
Programming costs	37%	33%
Personnel costs	43%	32%

Apart from this systematic externalization, venues can hire freelancers to reduce personnel costs. Despite the differences in the social protection systems of each country, the volume of labour performed by external personnel is remarkably similar: 18% of worked hours are done by externalized personnel in French venues, whereas 20% of work volume is performed by externalized workers in the Netherlands. These freelance workers are mainly technicians, as a large share of artists' employment is included in programming costs paid to external organisations who are in charge of paying the artists. Venues in both countries tend to prefer the most protected status for their workers: 88% of externalized work is done by *intermittent* workers in France (with 12% on 0-hour contracts), while 56% of externalized work is done by 0-hour contract in the Netherlands, the rest being done by freelancers. Finally, a difference exists in the proportions of employees on the venues' payroll, which is slightly higher in France (70%) than in the Netherlands (62%). This difference is explained by a higher use of interns in the Netherlands, which represent 5% of French venues' workforce and 15% of Dutch venues' workforce (Table 2.4).

Table 2.4 Distribution of working hours by type of employment. *Source* VNPF (Annual Survey 2016) & FEDELIMA (Observation Participative et Partagée 2016)

	France	The Netherlands
On venue's payroll	70%	62%
Freelance, 0-hour, *intermittent*	18%	20%
Internships	5%	15%

Hence, whether Dutch or French, pop music venues rely heavily on an externalized workforce in order to reduce their personnel costs. They assume mainly the diffusion part of concert production and thus, as opposed to the venues studied by Baumol and Bowen, *de facto* externalize the content production costs onto specific organisations or musicians themselves. They only assume the costs linked to the production and promotion of events in and potentially outside their walls. In order to pursue their activity, they rely on freelancers, hired when the personnel needs increase, typically when a concert happens. They rely on this type of personnel in similar proportions, despite the large differences between the respective policies concerning the social security of freelancers.

The De-commodification of Freelancers' Labour Market as an Answer to the Effects of Baumol's Law

It is clear that in their current form, live pop music venues need public funding to sustain their activities both in France and in the Netherlands. Even though a few venues situated in the centre of large and cosmopolitan cities are able to produce a profit, most venues, which ensure that the cultural offering is fairly distributed across the country, would not survive without public funding. This undermines the doubts of Cowen (1996) on the applicability of Baumol's law to popular music. Personnel costs represent the largest cost in both countries, although the production of the staged spectacles is externalized, and that venues rely a lot in both countries on freelancers to function, which is the norm for hiring technicians.

Cutting personnel costs through externalization was not identified by Baumol and Bowen as a way to bridge the income gap resulting from the cost disease. Even though it constitutes a direct and efficient way to tackle rising personnel costs for venues, it comes at a social price. In the Netherlands, commodification of labour has the consequence that sound technicians' earnings are not defined by their needs, but by how much they can ask given their labour market position:

> "I have a minimum for myself. [...] it was basically based on a certain amount of hours that I would work in a month. And a minimum income that I would need to guarantee that I can pay my rent. [...] basically one or two guarantees that I could support my life, basically.
>
> Z: [...] And how much is this minimum?
>
> It's 20. 20 euros per hour. But that's excluding BTW. The client also pays the BTW. [...] I mean, it sounds like it's quite high, but you don't have any security as a freelancer of course. If you get sick, or something, you don't get any money. [In fact], 20 is definitely on the low side. I know people who work for like...16, 15 sometimes. But that's really, it's not enough for the freelance work. But I think the average for sound technicians, especially in clubs is basically 25. Maybe it's 22,50. Around 25. And that's, I think that's global.
>
> Z: And you negotiate the price client by client?

Yeah. It really depends, you know, because some places are so small. But then for certain nights they do want sound technicians. But it's like...maybe really big name DJ that's playing there and they want to make sure that if something goes wrong there's someone who can fix it." (NK, interview)

NK is a relatively young technician, who has a 2 years' experience as a permanent venue technician, and 6 months as a freelancer at the time of the interview. He is trying to not go below a necessary minimum in order to make a living by working. However, he is mentioning that the market conditions can pressure him to go below that minimum, and that some people actually accept these conditions. This tension and uncertainty about how much a job will pay is also found at the other end of the income range, as this respondent claiming to earn around 80 000 € a year describes:

"Z: How do you negotiate your fares? How do you decide on the price?

For me, it's been a little bit middled to what in a certain way is reasonable, and also that you are not being...becoming too, er...more expensive than what's kinda like the average price. Because what then will happen is the only thing that happens then, is, they only gonna call you for, to do the show.[...] if I make myself too expensive, they're not gonna call me for the load in, because there's tons of guys that work for cheaper than me. So I am looking at, okay, what is kinda like the normal rates. I am a little bit above, but not so much that I am not interesting anymore to call for a load in or load out. And this actually helps me that I then have, let's say, on a weekly basis, let's say four working days out of seven. And then there's enough income for me to, yeah, live. Live my life and do my things." (NA, interview)

This tension is due to the absence of a stable employment relationship: the risk of job loss is always present, and freelancers have to develop their own strategy to counter the risk of loss of activity. This loss can be conjunctural, resulting from the financial failure of employers, but it can also be health related. No system of public insurance ensures freelancers, who have to earn enough in order to pay private insurance. And, as we saw, new entrants into this labour market tend to barely reach a minimum living wage, younger interviewees generally do not have better protection than just being conscious of their health:

"You don't have any security as a freelancer of course. If you get sick, or something, you don't get any money." (NK, Interview)

"Z: As a freelancer, you...you, if something happens, I don't know, you break your arm or something like that

You're fucked [...]. If anything happens with me right now, I am just screwed. And I wouldn't have anything...

Z: Knock on wood

Be careful with my legs. And arms." (NB, Interview)

Private health insurance ensuring a health leave is considered too expensive by most respondents. It is only used by the ones earning more than 50 000 euros a year. Private local support systems exist to cover health-related risks, but they have a limited capacity and are thus only accessible to people who have already secured a position in the field.

"There's a thing in Holland called het broodfonds, which literally means a bread fund. It's actually exactly what you're saying now it's a social protection system in the sense that, it's an official setup, and the Triodos Bank supports it. And it's an unemployment insurance. but done socially so if...

Z: For freelancers, you mean?

Yeah. For freelancers. So I am in one now in Utrecht. And it needs to be at least 25 people, maximum 50. This is statistics. The risk is calculated and stuff. But...yeah, now basically I pay a certain amount of money per month. I think it's 55 euros. And then there's different levels of...return that if I fall ill, I receive an extra amount of money, from the other 25 who are part of the gang. And so when someone falls ill, everyone gives a part of their money." (NI, Interview)

Conversely, in France, almost all respondents are employed in the regime of freelancers in performing arts. It means that they benefit from public health insurance like any other employee, and that they contribute to public retirement insurance. Moreover, if they attain the threshold of 507 worked hours in a year, they secure an income for the next year. Most French respondents experienced few or no breaks in this status during their careers, except at the beginning of it. Although some side effects were mentioned by respondents, notably the fact that the intermittent status creates a hierarchy within the profession, respondents were generally positive about it. This goes in in line with what Menger (2015) noticed, that social conflicts about this specific regime were the only case known where employers and employees stood together. In this case, the sides are not employers versus employees, but employers *and* employees of performing arts, versus employers from other economic sectors.

From 2003 to 2016, the unemployment regime of freelancers in performing arts was the subject of a recurring social crisis. French public unemployment insurance functions through contributions to funds associated with trades. Schematically, a salaried worker employed in, for example, the construction sector contributes to the construction workers fund. This fund is managed by a joint committee of employers and employees, and transfers compensation to the unemployed workers of the sector who have contributed enough. Hence, whenever a sector enjoys full employment, contributions exceed compensation and funds are in surplus. Conversely, when a sector suffers from a lack of labour demand and there are numerous unemployed workers, the compensations paid out exceed the contributions coming in, and the fund will be in deficit. This is where the principle of "interprofessional solidarity" enters the stage: deficits of certain funds are compensated by the transfer of the surplus from others. The terms of this "solidarity" are re-negotiated every two years by representatives of all funds, overseen by the state.

The cause of the recurring social crisis comes from the systematic deficit of the fund for freelancers in the performing arts (Menger, 2015). The terms of interprofessional solidarity are negotiated by the different representatives of each sector every two years, in order to fit the current economic situation. Representatives of wealthy sectors in manufacturing and services blame the structural deficit of the *intermittents'* fund, and press for enforcement of the conditions of access to the specific unemployment insurance regime. The latter do not accept the conditions offered, the state becomes a mediator, and a compromise is found potentially after a period of strikes in the cultural and media sector.

The 2016 compromise, obtained during a period of social agitation, solved the crisis by restoring the 2003 access conditions, and by involving the state's budget in counteracting the regime's deficit. However, I would argue that given that venues tackle rising personnel costs through externalization, the systematic negative balance of the regime contributions and compensations is not surprising. More and more freelancers are hired due to a growth in activity and the need to handle the effects of the cost disease. As compensations tend to grow faster than contributions (Menger, 2015), the fund can only be in deficit. However, what is interesting with the mechanism of interprofessional solidarity, is that sectors that benefit from technological productivity gains, and use them in a way that raises the general cost of the workforce, actually redress the balance by funding the regime that provides the features of a stable employment relationship to freelancers in the performing arts.

Since the 1980s, studies in the field of economic sociology have shown the social and cultural embeddedness of economic markets, and deconstructed widespread assumptions about the economy (Granovetter, 1985; Zukin & DiMaggio, 1990). Under the mainstream economics paradigm within which Baumol and

Bowen framed the cost disease, the concern of actors who bail out deficits might be justified. Indeed, if economic actors are first and foremost interested in profit, using productivity gains to raise production volumes and thus benefits is the obvious way to act. But if one takes one step back, and does not assume it is normal to convert productivity gains into production and benefit increases, the cost disease tells another story: the one of an economic sector unable to reach economic sustainability due to the production practices of another sector. In this perspective, sectors with high productivity gains are actually creating the cost disease in performing arts. Therefore, the transfer implied by interprofessional solidarity can be interpreted as a *compensation* of industry and services whose practices caused the "cost disease". It can be framed as a form of "polluter-pays" mechanism, where it is not the state or private funds who bail out performing arts, but the people who orient their activity for maximum benefit and in a way provoke the problematic economic dynamics leading to the income gap. This system of compensation, interestingly, preserves the autonomy of performing arts, while subsidies imply a form of control by the funder, as they are conditional on a defined set of venue activities. In turn, performing arts venues hire freelancers when they need them, and part of the cost of maintaining a decent level of income is covered by wealthy sectors initiating the cost disease.

Conclusion

In sum, the French response to commodification of work for freelancers in the performing arts constitutes also an answer, different from subsidies, to the effects of the cost disease. It is a way to reduce the financial pressure on musical performances. Venues can lower their personnel costs by externalizing personnel, which is something they mainly do with artists and technical intermediaries. But, with the system of *intermittence du spectacle*, the personnel alone do not bear the cost of such savings. As Collins (2004) mentions, one gathers the material means necessary to stage an interaction ritual only if one foresees sufficient reward in emotional energy. In this equation, sound technicians, and more generally technical intermediaries employed as freelancers, are an adjustable variable. Venues lower the material costs of the interaction rituals that are musical performances by externalizing their workforce. However, in order to engage in their job, technicians have to be sure of a guaranteed minimum material condition, otherwise even the most passionate people would quit.

The ability to provide the minimum that would maintain technicians invested their profession, central in Baumol and Bowen's concept of "cost disease", is an element of what I described in Chapter 1 as the "technical ability" of a group producing a music performance. The French compensation system forms a way to secure these minimum conditions, and, as we will see in Chapter 3, influences careers and the structure of relationships in the workplace. This latter way to tackle the cost disease should be further explored. We should more precisely quantify the range of the compensation. This requires a measurement of the increase of personnel costs and fund transfers from interprofessional solidarity in a longitudinal study. Another interesting focus would be on the effects of a limit on production increase as a result of productivity gains. Such a study would need to look for the economic opportunities such a move would create in performing arts, or other sectors which, due to the nature of their activity, cannot catch up with productivity increases of industry and services.

Creative Craft in Music: Careers and Roles of Sound Technicians

3

In Chapter 2, we saw how performances are embedded in a global economic dynamic leading to externalizing the work of sound technicians, resulting in the scarcity of standard employment relationships in their labour market. We have seen that the project-based structure of their work, and the systems of social security from which they potentially benefit, heavily shape their standard of living. However, although a project-based organization of work is a way to cut the economic costs of personnel and can lead to workers' precarization, it also has the effect of increasing the number of professional interactions within the field. Indeed, as people constantly meet new colleagues within different "project teams" (Hesmondhalgh, 2006), they are confronted with a diversity of methods, habits, ways of doing, ways of being. This constant blending has effects on the emergence of new practices in art worlds, leading to the emergence of successful and innovative cultural performances (Crossley, 2015; Patriotta & Hirsch, 2016; Uzzi & Spiro, 2005).

The effects of a project-based organization of art worlds on artistic careers and the consequences on aesthetic innovations are well documented. However, while the labour market of sound technicians is structured in the same way, the effect of this organization on both their careers and the composition of their tasks is uncharted territory. In this Chapter, I will bring elements gathered from the analysis of the career paths and description of job experiences of French and Dutch respondents, which will shed light on the structure of technicians' careers, the labour division in their workplace, and the influence of institutional contexts on these two elements. This will lead to an understanding of how technical teams are

Electronic supplementary material The online version of this chapter (https://doi.org/10.1007/978-3-658-33029-3_3) contains supplementary material, which is available to authorized users.

© The Author(s), under exclusive license to Springer Fachmedien Wiesbaden GmbH, part of Springer Nature 2021
A. Battentier, *A Sociology of Sound Technicians*, Musik und Gesellschaft, https://doi.org/10.1007/978-3-658-33029-3_3

assembled, and how they practically implement the transformation of the performance's script into a physical object which will be the target of a mutual focus of attention during the performance. I will also discuss how sound technicians aesthetically contribute, as technical intermediaries, to the performance.

Theory: Temporary Teams of Producers

Apart from its economic aspects, project-based organization of work heavily shapes the careers and lifestyles of artists. Career advancement and achievement are never sure, and artists need to get involved in several jobs related to artistic activities, developing a "portfolio career" (Menger, 1999; Perrenoud & Bataille, 2017; Throsby & Zednik, 2011). Social skills are also crucial, as managing relations is essential to be hired to work on projects (Dowd & Pinheiro, 2013) and creating a social environment open to the development of artistic ideas (Montanari, Scapolan, & Gianecchini, 2016; Reilly, 2017). Ashford et al. (2018) have shown that perseverance, anticipation, resilience and image management are necessary behaviours to sustain a portfolio career. Also, one must be able to cope with strong and oscillating emotions related to both the uncertainties of the market and the high level of personal involvement in artistic activities. Individuals involved in such labour markets can struggle for recognition and/or income in a hypercompetitive context, which can lead to abusive employment practices (Fast, Örnebring, & Karlsson, 2016; Hesmondhalgh & Baker, 2010; Percival & Hesmondhalgh, 2014). Nevertheless, artists, as well as other actors involved in the making of performances, often accept these uncertainties by developing a vocational, passionate relationship towards their activity (Sapiro, 2007; Sorignet, 2010).

However, this type of organization has also been shown to be a way to fuel the circulation of ideas that entails successful performances, by offering the possibility of a balanced circulation of people, with the ideas, conventions and practices that they carry. Burt (2004) showed that individuals situated between groups, in what he called "structural holes" are more likely to develop creative and original ideas. Patriotta and Hirsch (2016) showed that "amphibious artists" occupying the structural holes between the core and the periphery of US movie industry played a central role in the emergence of the field of independent films. The notion of the amphibious artist can be embodied through groupings of individuals: Cattani and Ferriani (2008) have shown that the associations of people, whether they are artists or cultural intermediaries, respectively from the core and the periphery of the movie industry were more likely to produce well regarded films, virtually filling a structural hole by benefiting from the different advantages of both positions.

Finding ways to connect individuals with different visions, coming from different groups but still likely to share a common pool of conventions appears essential to artistic innovation, which is itself related to the possibility to produce re-fused performances. Uzzi and Spiro (2005) have shown the existence of a correlation between the "small-worldness" of an artistic network and its ability to be granted commercial and critical success. They described "small-worldness" as the tendency of a network to contain individuals that bridge between different clusters. They showed that the proportion of successful plays rise with the small-worldness of the network up to a critical point, after which it starts to decrease.

Practically, Uzzi and Spiro showed that members of an art world, if they collaborate too repeatedly, tend to homogenize their beliefs, behaviours and viewpoints. As a result, they homogenize their practices, and self-validation percolates into the group. People in this mindset become more and more reticent about creative innovations that could lead to originality and success. Individuals circulating between these groups break this dynamic by having different stances on reality, which make them come up with original and unintended ideas, nonetheless not too remote from each group's culture and thus likely to be validated by them. The project-based structure of the artistic labour market is a way to tackle this mechanism, and to blend people in a way that favours the production of performances that deliver both critical and commercial success.

Artists and technical intermediaries' careers share a similar pattern of succession of short-term collaborations. They both circulate between different groups, work on various projects, and are constantly confronted with more or less familiar colleagues. The question, then, is how the latter contribute to these dynamics of artistic innovation. In order to answer these questions, it is necessary to understand how technicians' trajectories are driven between various jobs, and how their different tasks are distributed.

In the workplace, artists and technical intermediaries deliver teams within temporary organizations. These teams are set up and dissolved according to the needs of each project, and therefore cannot rely on a structured, institutionalized and clearly identified hierarchy. Bechky (2006) has shown that these teams coordinate through a role structure. For instance, on a movie set, people are hired to perform a defined role, such as director, gaffer, production assistant, wardrobe supervisor. The bundle of tasks attached to their role (Hughes & Chapoulie, 1996) are broadly known by employees due to their previous experiences in similar positions, and is incrementally defined through social pressure in the form of intense thanking, admonishing and joking. With the help of this pressure, individuals involved both define and negotiate their role boundaries: they identify their expected tasks and behaviour, and can try to gain influence within the team.

In the crews Bechky studied, some roles are defined as "artistic", while other are defined as "technical". However, existing ethnographies of technical intermediaries show that this difference can be fuzzy. In music production, studio sound engineers often provide an artistic contribution to a record, in the sense that they take autonomous decisions that shape the record aesthetics (Kealy, 1979; Perrenoud, 2007; Rudent, 2008). Film editors* of TV news report can also have a relative autonomy in aesthetic choices if the topic is not considered important by the publication chiefs (Siracusa, 2000). Even on tasks that could be considered as "purely" technical, such as handling the logistics of a stage building, are potentially claimed to have an artistic dimension (Laborde, 2008). This distinction between the "artistic" and the "technical" is also very much present in the accounts that the informants give of their work in the case of this dissertation, with the same ambiguity.

In the remaining sections, I will focus on how sound technicians' careers are chronologically structured, from the decision to work in the field to their advanced professionalization. I will then untangle the distinction between "artistic" and "technical" tasks in their workplace. I will finally discuss the differences found in the two different institutional contexts and relate them to the discussions of Chapter 2.

Method: Thematic Analysis and QCA

Interviews are the best fieldwork material to study the career path and aesthetic nature of sound engineers' positions in music worlds. Indeed, they illuminate the different positions occupied by the respondents, both chronologically and by role (Interview grid in Appendix C). From this material, I will build a corpus of job experiences, characterized using different properties relevant to the questions asked in this chapter. This corpus will then be analysed using thematic analysis and qualitative comparative analysis.

All interviews were fully transcribed and coded with Atlas.ti, in two waves. First, I identified all work experiences mentioned by the respondents, and characterized them by working context (on tour, in venue, working with a service provider...), position (front of house, assistant, studio engineer...), and artistic genre (musical genre, theatre, cinema, television...). A three-level code also accounted for the importance of the work experience within the respondent career (anecdotal, significant, long-term project). A total of 315 work experiences were mentioned in the interviews. In each of these experiences, respondents had endorsed a role that was described according to the four coded characteristics. Hence, this first wave of coding accounted for terms used by the respondents to describe their job.

The second wave of coding transcribed these codes in theoretically relevant and empirically grounded analytical categories. Positions were gathered in function of the specific skills that they require, and the hierarchical position of the respondent. Artistic genres were gathered in four categories. "Non-artistic activities" correspond for instance to corporate events or working for an administrative institution. "Artistic but nonmusical" activities cover theatre, cinema and other. "Music genre indifferent" was applied when a respondent mentioned that his role was not tied to a particular genre of music, such as in the case of a venue sound engineer who welcomes different kinds of bands. Finally, "music genre specific" was applied when the respondent's role implied working within a given music genre.

"Context" corresponds to the type of institutional context in which the respondent was working during the given work experience. The distribution of the context variable was heavily concentrated on a few codes that correspond to typical working contexts of technicians in musical settings, whether in France or in the Netherlands. Hence these dominant codes were kept, while all the others were gathered in a miscellaneous category. However, given the importance of electronic dance music in the Dutch market, and the fact that many Dutch respondents mentioned some experience in clubs, the corresponding code was extracted from the "others" category and established as an analytical category despite not occurring as much as the other ones.

Finally, I accounted for the "artistic dimension" of the technician's role by evaluating from the respondent's tale whether his/her role implied the presence of autonomous decisions. Following a distinction between arts, arts-related, and non-arts jobs in the analysis of the portfolio career of artists (Menger, 1999; Throsby & Zednik, 2011), each role was coded to account for its aesthetic implications. A role could imply the "presence of autonomous aesthetic decisions", such as in the case of a front of house engineer who has to decide how to balance the instruments. A role can be "technical in an artistic context", such as in the case of an aide who essentially assists the main sound engineer by handling the stage installation. In this case, if no formally autonomous aesthetic decision is made, the technician however evolves in the specific context of an art world and needs to manage its conventions. Finally, a role can be either technical but in a non-artistic context, such as installing the PA system for a union's conference, or non-technical, such as teaching or writing in specialized magazines. These two latter cases were blended in a single category (Table 3.1).

Table 3.1 Summary of the coding process

First wave coding	#	Second wave coding	#	Code
Artistic dimension of the role				
		Non artistic context or non-technical position	23	AUT
		Presence of autonomous aesthetic decisions	165	TEC
		Technical work in an artistic context	128	NARA
Working context				
Boite de nuit	11	Club	11	CLU
Boite de presta	47	Service provider	47	SER
Festival	50	Festival	50	FES
Salle de spectacles	139	Venue	139	VEN
Studio	54	Studio	54	STUC
Tournée	115	Tour	115	TOU
Administration	1	Other	84	OTH
Association	10			
Audiovisuel	13			
Bar-Restaurant	5			
Boite d'intérim	3			
Captation	4			
Centre de vacances	2			
Compagnie de théâtre	8			
École de musique	7			
École de technicien	2			
Espace d'accueil d'artistes	1			
Fondation	1			
Freelance indépendant	5			
Institution international	1			
Label/Tourneur	10			
Lieu privé	4			
Musée	1			
Prod cinema	3			
Résidence	1			
Revue spécialisée	1			
Université	1			

(continued)

Table 3.1 (continued)

First wave coding	#	Second wave coding	#	Code
Genre of activity				
Balkanique	1	Music—Genre Specific	173	GSP
Chanson française	5			
Classique	28			
Cover Band	1			
Dance	10			
Electro	26			
Flamenco	1			
Funk	2			
Fusion	2			
Gospel	1			
Hip Hop	3			
Inconnu	17			
Instrumental	1			
Jazz	26			
Metal	2			
Musique contemporaine	1			
Pop	12			
Punk	3			
Reggae—Dub	2			
Rap	5			
Rock	7			
Rock français	6			
Ska	2			
Soul	1			
Traditionnel	2			
Trip Hop	1			
Tzigane	1			
Variété française	1			
World	3			

(continued)

Table 3.1 (continued)

First wave coding	#	Second wave coding	#	Code
Formes variées	103	Music—Genre Indifferent	104	GNS
Soirée privée	1			
Ciné-concert	1	Artistic non music	53	ANM
Cinéma	6			
Comédie musicale	2			
Installation	3			
Jeu vidéo	1			
Jeune public	1			
Production de vidéos	4			
Radio	6			
Réalité virtuelle	1			
Spectacle d'arena	5			
Télévision	4			
Théâtre	19			
Administratif	1	Non Artistic	23	NARG
Animation	2			
Audioguide	1			
Enseignement	4			
Escape room	1			
Évènementiel	13			
Expérimentation sociale	1			
Position				
Assistant	51	Aide	158	AID
Assistant capta	1			
Assistant lumière	5			
Assistant plateau	11			
Assistant son	12			
Roadie	23			
Stagiaire	55			
Accordeur de piano	1	Backline	8	BAC
Backline	7			
Design sonore	16	Creation	16	CRE
Ingé capta	22	Sound for image or broadcast	22	IMA

(continued)

Table 3.1 (continued)

First wave coding	#	Second wave coding	#	Code
Chef d'entreprise	3	Management	44	MAN
Directeur artistique	2			
Directeur technique	2			
Gestion d'un lieu	3			
Régie générale	34			
Musicien	10	Musician	11	MUS
Arrangeur	1			
Ingé accueil	72	PA	260	PAS
Ingé façade	97			
Ingé façade (conduite)	14			
Ingé retour	73			
Poste tournant	1			
Référent son	1			
Responsable HF	2			
Ingé studio	36	Studio	44	STUP
Mastering	1			
Post-production	7			
Conception de systèmes	1	System	25	SYS
Ingé système	24			
Formateur	5	Teaching	6	TEA
Rédacteur	1			
Régie lumière	33	Technical but not sound	41	NMU
Régie plateau	7			
Régie vidéo	1			

This systematically described corpus of work situations provides the empirical basis for outlining the career paths of technical intermediaries, as well as the on-the-spot hierarchies and their relationship with the aesthetic contribution of technicians.

In order to answer the question of how the careers of technical intermediaries are structured, career path analysis, using a sequence analysis, is theoretically the best solution (Abbott & Tsay, 2000). However, it requires a significant amount of structured and quantitative data. In the case of technical intermediaries, such data do not exist, and there is no previous work from which it would be possible to draft a methodology. Indeed, existing studies of technical intermediaries mainly study their role on the workplace, but not their career path. Hence, the career structure of technical intermediaries has to be approached qualitatively, through a thematic analysis (Braun & Clarke, 2006; Nowell, Norris, White, & Moules, 2017). This approach will allow the identification of the conceptual categories relevant for a study of technical intermediaries' career, and will provide insights in their structure and progression over time.

In order to account for technical intermediaries hierarchies in creative work, I proceeded to an analysis of sufficiency via a crisp-set qualitative comparative analysis (QCA) (Ragin, 2008; Rihoux & Ragin, 2009; Schneider & Wagemann, 2012), with the code "artistic dimension" as an outcome, and position, context and genre as conditions. The purpose of this analysis is to understand how technical intermediaries are involved in the construction of meaning according to the role they are given in temporary project organizations, by successively asking which positions, contexts, and genre tend to offer more or less aesthetically influential roles to technicians.

The findings presented in the two next sections are common to France and the Netherlands. In a third section, I will focus on the differences found between the two countries, and interpret them through the prism of the institutional differences exposed in Chapter 2.

The Career Structure of Technicians: Vocational Impulses, Education, and Constellations of Jobs

The first finding that can be highlighted in the corpus of work experiences is the confirmation that like artists, technicians work on a project-basis and develop a portfolio career. The quasi-totality of work experiences of the corpus are short-term collaborations that can be repeated over time, but with no formal guarantee of re-employment. This implies that technicians are subject to pressures related to job uncertainty similar to the ones applying to artists. Similar to artists, developing a passionate relationship to their profession is a way to cope with these uncertainties. Indeed, all respondents, whether French or Dutch, declare a strong attraction towards music activities, often dating back to their childhood. The vast majority of respondents had mastered a musical instrument, and many maintained their proficiency. A few of them added professional musician to their technical activities, although most respondents mention strong difficulties in combining both activities. Indeed, the role of artist and the role of technician require different abilities that require time to develop and maintain, including the development of distinct professional networks for job finding. Hence, one generally needs to choose his/her side of the stage:

"I was interested into playing with people. Playing in quartet, orchestra... And with my planning it was impossible to ensure a rehearsal, each Tuesday, 6.30... Half the time, I was ditching people. So that didn't work well. Moreover, in my training, I learned to... I learned sound, I developed my ear, the instrument's sound, all that, all that. And the violin is pretty ungrateful. When you come back, you take your whatsit and it's been six months you haven't play, well, it doesn't sound good." (FP, Interview)

"I use to play drums and guitar. But I kind of stopped doing that when I started doing all the technical stuff. And now I kind of produce music. But on hardware instruments and a computer. But I... completely stopped playing instruments. Which I sometimes really regret." (NK, Interview)

Once this attraction towards music is accounted for, is the choice for the shadow of backstage a choice made by passively, becoming a technical intermediary being the dead-end street for failed musicians? This is clearly not the case: many respondents are highly musically trained and chose to work as technicians despite having the training and the opportunities to become professional musicians. Furthermore, the majority of respondents expressed their preference for technical tasks. Some respondents do not like to be in the public eye and prefer to work in the discretion and anonymity of technical positions, rather than being under the spotlights. The specificities of technical positions are also mentioned as advantages of technical roles. The control granted on the performance's output, the constant search for a compromise are often mentioned as enjoyable features of the job, as well as attraction towards the tools of sound reproduction technologies (mixing desks, PA systems etc.). Technical intermediaries' roles thus count a number of specific features that motivate individuals to actively choose the shadow of the backstage.

The call towards sound engineering can be an early one: some respondents determined their career choice very young and oriented their education towards this goal. Another current profile is a reorientation after a disappointment in another, more "classic" profession. In this case, the engagement in sound engineering profession was postponed due to the social pressure towards a "normal job", perceived as more stable and financially secure.

"I had this project [to become a sound engineer] since I was 16 in fact.

Z: You didn't start...

No, because around me, I had talked a lot about it, but I have been told "no, it's too complicated...it's too risky, you won't want it in the end" so I had been somehow...I never launched the process. So then, this skills assessment arrives. And people tell me again, the same speech, saying "careful, very complicated, you won't live of it, it's not reliable, it's difficult..." [...]. So, this option is out at this moment." (FG, Interview)

Flocco and Vallée (2012) perceived that the familial environment plays an important role in one's engagement in the profession of movie machinists, the technical intermediaries that build movie sets structure and handle camera movements. My data show that this is also the case in sound engineering professions. Siblings accustomed to the practices of professional art worlds provide emotional support, information on the appropriate behaviour in the labour market, and first opportunities: many respondents mention a sibling mentoring or providing through his personal network the first contacts with the field, sometimes when the respondent was still a teenager. On the contrary, those who do not have a parent involved in art worlds, whether as an artist or a technician, encounter difficulties in finding their first work experiences, as well as to understand the codes of the profession.

A phenomenon of path-dependency (Bernhardt, 2001) seems to exist in a sound technicians career. Indeed, the first working experiences appear to determine significantly the career path of respondents. They tend to "snowball" their next work opportunities and progression from these first experiences. They thus develop a competence specific to a working context, a position, and/or an aesthetic. They can use this competence to "jump" to another position, context and aesthetic using a transferable skill, and start working and developing a network of working contacts in another art world. Hence, while artists careers can be considered as cumulative or "layered" by their several achievements within one art world (Reilly, 2017), sound technicians career paths look more like "constellations" of trajectories within different art worlds. If they can climb the ladder of positions of responsibility by deepening their involvement within an art world, their evolution goes partly sideways as they integrate different art worlds using transferable skills.

For instance, FO started doing sound engineering with his group of musician friends in the Paris suburbs. He decided to train as a technician, and started in a specialized school. He did two internships during his studies. One was at a jazz club in Paris, another one in a web radio. He started to work in both places after his internships and developed both activities in parallel, sometimes grabbing the opportunity to work in another jazz club, which is in the same area. These two positions now provide most of his working hours, but he had a period of touring as a backliner*, a position offered by a drummer who played at the club, whose band was performing long sets for luxury private parties. Another example is FQ, who started to volunteer each year at a jazz festival* organized by his uncle when he was a teenager. After 2 years of technical education in audiovisual technologies in the North of France, he came back to the West where he comes from, but did not find enough work. He joined his girlfriend in Paris, who was working as a video technician and introduced him to work on TV sets. He managed to get a liveable income within 6 months. After this, in concerts he attended as an audience member in Paris, he met technicians that knew him from the time he was working in his uncle's festival*. They offered him work with a service provider they were working with, to which he had already applied but got no positive answer as he was not recommended. He then started to work in the music field again, and little by little got specialized in system calibration. This expertise stimulated his recruitment as a system engineer by touring companies for nationally renewed artists. However, FQ did not stop working in television. Indeed, touring is exotic and exciting but also tiring and difficult to reconcile with family life. Hence, he alternates between both activities to maintain a satisfying work-life balance, with periods of more intense involvement in one of them.

Due to this "constellation" structure of their career, transferability of skills from one world to another is essential to the respondents' career development. Transferability is eased by the presence of a formal education in these skills. Respondents ascribe this to the expansion of formal education among newcomers, contrarily the profession used to be mostly learnt on the spot during 20th century. Education is generally perceived as useful, potentially necessary, but systematically not sufficient. This is in line with Bechky's (2006) account of the centrality

of role definition and embodiment within temporary organizations: technical intermediaries' theoretical skills can be learned by a formal education, but they are
improved and developed along professional evolution, and the boundaries of the
role one shall fulfil are learned through experience on the spot. Respondents'
feeling towards their formal training range from perfectly useless to extremely
useful. The worst impression has been left by Dutch private/public training in a
technical work.

*"There was one teacher for sound. That's how bad it was actually. [Laugh] And he was,
this guy, like failed to get work. Because he just was not good, because people didn't
want him. And...Yeah, he had like two or three microphones and he was showing...stuff.
There was this recording...He let us listen to a recording once. He was very proud of
this like 'yeah, this is my band and I recorded this and bla bla bla'...I listened to it it
sounded like shit..." (NC, Interview)*

Private schools as well as French short technical training[1], delivering a diploma in
2 or 3 years, leave the impression of a superficial overview where one learns the
very basic skills to start on the job. Long terms studies (5 years), only present in
France, are very positively perceived by the trainees, although they still mention
the necessity to challenge their theoretical knowledge with the "reality" of the
field. However, all trainings orient their courses towards a lot of practice and
internships.

Becoming a technician is the result of strong career choices. Positions are hard
to access and starting a career in the field is eased by the presence of artists and/or
technicians in one's circles of primary socialization. If the job can be learned on
the spot, following a cycle of formal education provides a general background
knowledge that improves one's ability to be employed in different positions, contexts of work or aesthetics. Besides, education provides internship opportunities,
that are pivotal for the integration in professional networks. Indeed, the first working experiences shapes a lot the future career of individuals, as they are starting
points on which similar experiences will build with time. Transferability of skills
from one art world to another is a major difference between technical intermediaries and artists. Indeed, the latter tend to accumulate artistic, art-related and
non-artistic work in the art world in which they are engaged, while the former
can work in different art worlds according to their preferences, skills, and opportunities. I will now review the extent to which their investment in these art worlds
is more or less marked by aesthetic autonomy.

---

[1]*Brevet de Technicien Supérieur (BTS) Audiovisuel*

In Musical Performances, There is No Such Thing as a Purely Technical Task

In their work experiences, technical intermediaries can have roles that are more or less charged with autonomy, responsibility, and aesthetic leeway. What are the roles which are more granted with such features, and what drives their distribution within temporary teams? In order to answer these questions, I will identify which positions, contexts, or music genre are more likely to imply a role that is artistic (code "Presence of autonomous aesthetic decisions"), technical in an artistic context (code of the same name), or non-technical/non artistic (code "non-artistic context or non-technical position"). In order to do so, I proceed to a qualitative comparative analysis (QCA) of sufficiency of the contexts, positions and genres on the outcome "artistic dimension" (Ragin, 2008; Rihoux & Ragin, 2009; Schneider & Wagemann, 2012). Using QCA will allow the identification of which positions, contexts, and genres are most associated within the corpus of work experiences, which are the studied cases, with the three possible states of the outcome.

In QCA, a relationship of sufficiency between a condition and an outcome is assessed by the consistency score of the set of conditions in the truth table. In a crisp-set analysis, where membership within a set is binary, consistency score is the proportion of cases matching a configuration of conditions that also display a positive score in the outcome. For instance, if the outcome "technical in an artistic context" is present in all the cases in which the position "aide" is present, then this position will have a consistency score of one on the outcome, meaning that the set of cases with the position "aide" is fully included in the set of cases with the outcome "technical in artistic context". In such a case, I can conclude that working in an aide position is sufficient to work in a position that is technical in an artistic context. Empirically, it means that whenever a respondent is employed as an aide, the job will be technical in an artistic context.

In this study, I use consistency scores as "coefficients of association" that show the extent to which contexts, positions, and genres are associated with three types of more or less aesthetically engaged technical roles. Conventionally, I use the term "strictly associated" when a condition has a consistency score of 1 for an outcome, meaning that all cases with this condition display the tested outcome. I also use the term "systematically associated" when a condition has a consistency score higher or equal to 0.75, meaning that at least three quarters of the cases with this condition also display the tested outcome.

The analysis on the whole set of cases showed that jobs with autonomous aesthetic decisions are associated with positions, genres and contexts implying a close connection between artists and technicians. However, the data also showed the ambivalence of technical jobs within art worlds, implying that these positions always include a significant share of creative insights. The most important results are presented in tables in the following sections, but all detailed truth tables can be found in Appendix D and E.

"Artistic" Jobs

Sound design* and studio positions are systematically associated with the presence of aesthetic decisions (Table 3.2), which means that these positions are most of the time held in jobs where an important aesthetic contribution is made by the technician.

Table 3.2 Consistency scores of positions on the outcome "presence of autonomous aesthetic decisions", full set

	Consistency score	Number of cases
Aide	0.110	91
Backline	0	2
Creation	0.929	14
Sound for image of broadcast	0.600	15
Management	0.615	13
Musician	1	8
Studio	0.912	34
PA	0.714	119
System	0.222	9
Teacher	0	5
Technical but not sound	0.400	5

Both positions imply a form of long-term work, with a lot of exchange of views between artists and technicians. The position "sound design" is the one with the most creative implication from technicians. In this case, technicians make direct aesthetic propositions and sometimes directly compose and record them. Sound design* is a position present in many different art worlds: a respondent holds one in radio, two in recorded music, respectively classical and electro music, two are designing the stage sound for bands they work with, one is creating soundtracks in theatrical plays, and another one in the design of escape games in Amsterdam. Sound design* consists of materially implementing an idea that has never taken any material form. Thus, there is no previous basis to compare the current work with a former one. Working within a clearly identified aesthetic category, such as genre, is a way to provide common grounds to the technical–artistic collaboration. Hence, genre-specificity is also systematically associated with jobs in which technicians have artistic leeway (Table 3.3).

Table 3.3 Consistency scores of genres on the outcome "presence of autonomous aesthetic decisions", full set

	Consistency score	Number of cases
Genre specific	0.753	150
Genre indifferent	0.355	107
Artistic but non music	0.447	38
Non artistic	0	20

These positions require a form of intimate collaboration and smooth communication with the artists. The project-based organization is useful to this dynamic, as it allows us to test whether a fruitful collaboration relationship can exist between the artist and the technician:

"I never really decline offers, I always redirect, like, if someone comes to me and wants to do an album. Because they like the other works, that's a great honor, you know. But sometimes I can already feel like that it won't match or…either, you know…they want to work in a way that I really don't agree with…something like that. Then I'll just say 'you know, I know a lot of other guys, a lot of other studios, what don't you have a talk with them, see if you can find a match'. So, for me it's always about, the right match. And… I guess I've sort of developed a moral compass in finding the right people to work with." (NI, Interview)

"[I like] the fact that I can change, all the time, of universe. It's something that sounds risky when you don't know it but well…when there's some work, jobs come one after the other. And the fact that you can change of place, not always see the same people. Especially when sometimes you can…you can have some chafing with someone sometimes it's better to not work with this person for a few months. You will meet them later, cross them […]. And eventually say, okay, we don't belong so take someone else next time, that's okay. I'll find something else, everyone finds something else, and everyone goes on. No one has to carry a millstone." (FF, Interview)

Kealy (1979) noted that once this type of work relationship has been developed, it is often necessary to follow up on it. Typically, a technician who got deeply aesthetically involved in the recording of an album is likely to go on tour with the band to handle the front of house sound. This is visible in my data as well: touring is, with studio, a context systematically associated with the presence of aesthetic decisions (Table 3.4). If a specific work of sound design* has been made, and the technician involved in the creation cannot tour with the performance, a work of transmission has to be made:

> *"And at the same time that I worked on this I was also doing concert with small bands, and eventually other shows on which I was working. But it was mostly as a* regisseur, *which means that I wasn't making the soundtrack, I was taking the show as it was given to me and I was applying the sequence...*
>
> Z: *Ah ok, you were not on the show's creation...*
>
> *I was, but as an assistant."* (FF, Interview)

On the other hand, working with a service provider, which rents out personnel and gear to paying customers, is the context least associated with jobs implying autonomous aesthetic decisions, with a consistency score of almost 0 (Table 3.4). It means that when a technician is hired by a service provider, he/she will almost never be granted the responsibility to make autonomous aesthetic propositions. This is logical as in such a context, artists and technicians generally do not know each other before the performance, which implies a certain caution, even a restraint, in their aesthetic initiatives. However, technicians employed by a service provider are often "welcoming technicians": they finalise the sound system and pass the baton to someone that will mix the orchestra on stage.

Table 3.4 Consistency scores of context on the outcome "presence of autonomous aesthetic decisions", full set

	Consistency score	Number of cases
Festival	0.483	29
Service provider	0.094	32
Studio	0.827	52
Tour	0.861	72
Venue	0.262	61
Club	0.636	11
Other	0.397	58

"Technical" Jobs

Within a job, several tasks are more technical, in the sense that they do not require one to directly act on the content produced by artists, and are hence not imbued with much influence on the aesthetics of the final object. These kinds of tasks consist for instance, as we have seen in Chapter 1, of setting up the amplification system for the night. The main mixer can also be given with the help of various aides in order to stay focused on the specific task of mixing: stagehands*, roadies*, microphone specialists. If these kinds of positions do have less aesthetics influence on the final object, the more or less complex and urgent technical matters they deal with are generally derived from the art world conventions, and people who are granted these positions have to properly understand them.

In fact, no position, context, or genre appears to be strictly associated with technical work performed in an artistic context[2] (Truth tables in Appendix D.2). This result deconstructs the idea of the existence of purely technical tasks during a creative process. Indeed, in an art world, even logistics tasks imply a form of creative composition. Laborde (2008) mentioned that "two concerts" were happening during the setting of an opera. The first is the one that the audience will see, while the second is the coordinated actions of stage builders, which the stage manager* is the *chef*.

FC best exemplifies the aesthetic influence of apparently purely technical tasks. Despite possessing one of the highest diplomas for sound engineering in the country, she generally prefers to work as an aide:

> *"I think I am one of the rare people from my school who likes to be an assistant. And who does not have the ambition to become chief. […] I love assisting, and quickly analyze what my chiefs need. And do the best so that their work is simpler, quicker, and more efficient. It depends on people, but I love this sociological analysis part of people with who I work. So, it's also for this that I don't necessarily try to make more…projects as responsible, and that I try to keep this aspect…simplifying the life of the person with who I work."* (FC, Interview)

This taste for the social and problem-solving aspects of her profession, combined with the advanced training she received leads her to be recruited frequently as an effective all-rounder on many of her gigs, which often appear to be quite exotic and experimental:

---

[2]Backline positions are technically strictly associated with technical positions, but this result is obtained on only two positions.

"The day that was halfway, in fact they decided to do something slightly new. Doing a concert, no, that was still filmed but that was filmed 360 degrees, with 360 degrees cameras. And the recording also had to be able to...how can I say...restitute* the sound on 360 degrees if you follow the camera. The idea was that you, spectator, can hear the concert with a helmet as if you were the camera. And to move around in the orchestra, visually seeing it around you and that the sound moves at the same time that you move your head when you hear sound in the orchestra. [...] So we installed the orchestra in some kind of circle with a travelling in the middle and a 360 camera above. That was moving vertically as well, anyway it was a mess to settle also for the video team, it was great. And us, we had suspended, well did a different installation, of wireless microphones on the singers because there were soloists. I took care of the soloists as well, once again for the wireless mics." (FC, Interview)

In her jobs, FC is usually one of the cornerstones of the technical team, solving problems and coordinating by discrete and numerous activities situations that might otherwise result in confusion, consuming time and energy for the rest of the team. Obtaining this result is a matter of anticipation, both of potential technical issues, but also a knowledge of what the other people involved in the performance production need. To effectively do so, people need to carry an intimate knowledge of the conventions, both aesthetic and relational, of the art world they work in. Finding the ways to ease the coordination requires one to creatively build on these conventions, and thus taking part, although potentially modestly, in the final shape of the music. For instance, when she was hired to set up the wireless microphone system for operas, she quickly understood the necessity of a good relationship with the wardrobe assistants:

"For me it's a new environment, costumers. And it's really crazy because you actually have to get friend with them. So that the hairdresser accepts you put a microphone under her beautiful thing...that the costumer accepts that you sow some stuff...well...and in fact it's really, really...for me it's 75% relational and 25% of technical mastery really, to do wireless microphones. Once people have seen that things were going well relationally you won...you did the job, almost." (FC, Interview)

Even though she is archetypal of the creative leeway of aides, and that her taste for assisting is rather unusual however, this idea that one must find creative solutions when apparently performing non-creative tasks is found in all interviews. Relational work, in sum, has to be construed as an object of creative work informed by sensibility towards the artistic matters of the performance.

Some contexts appear more associated with technical roles than others (Table 3.5). They follow the logic outlined in Table 3.5: working in a venue or for a service provider is more likely to imply a technical rather than aesthetic role, although this does not appear as a very clear-cut relationship in the truth tables. The technician employed by a venue or a service provider is more likely to be a "welcoming technician", not directly involved in the shaping of the music.

Table 3.5 Consistency scores of context on the outcome "technical role in an artistic context", full set

	Consistency score	Number of cases
Festival	0.517	29
Service provider	0.656	32
Studio	0.154	52
Tour	0.139	72
Venue	0.721	61
Club	0.364	11
Other	0.397	58

Non-technical Jobs and Technical Jobs in Non-artistic Contexts

The respondents were selected on the criteria that they were technicians working at least partly in music worlds. Hence, logically, much less non-technical or non-artistic work experiences are present in the interviews, but I cannot deduce that technical intermediaries in general are attracted to artistic applications of their skills. In this group of respondents, working in a non-artistic context is generally perceived as an accident. If the truth tables (see Appendix D.3) results are heavily biased by the process of respondents' selection, it is possible to distinguish two types of profiles in non-artistic contexts or non-technical positions. The position "teacher", which groups together teachers and people who write articles in specialized revues, includes professionals who find a way to transmit their knowledge and experience of the field through these peripheral activities. It is perceived as an appreciated extra activity, often mentioned in the end of the interview when I was confirming that nothing had been forgotten. The other profile is technical work in a non-artistic context. This type of job is rare in the whole panel, as respondents generally tend to avoid it. It is described as thankless and potentially badly paid:

"The [venue welcoming various corporate and politic events] I did it at the beginning to make money but [...] it didn't really fit me. [...] It's really some sort of conference, some, I don't know, Bouygues Telecom, *McDonald's or the Socialist Party that came and did their speeches. You put two swan-neck microphone and you listen the mister*

talk. Well it wasn't music and...I wasn't interested by the substance of it. [...] That was typically a job to put food on the table." (FC, Interview)

"[I worked for] a company that sends people to festivals* and... assist other companies when they need more hands to build a stage, to build a PA, to break down [...] Really bad experience, really bad...really heavy work, really seriously underpaid, and really dangerous, you know. Physically harsh...like 6 euros per hour. Something ridiculous [...] in Amsterdam. Really, really few money for the...for how difficult the job was, you know..." (NF, Interview)

The method chosen to collect and analyse data puts this third type of work out of the scope of my analysis. However, we can see that respondents tend to accept jobs that are technical in non-artistic context by default, as a means of subsistence, and exit this type of position as soon as they are granted the opportunity to do so. Some respondents also teach and write in journals as an extra activity. Conversely, they tend to like this type of activity although they consider it as remote from the core of what they do.

Nationally Bounded Institutional Differences: Education Programs and Access to "artistic" Positions

Differences in Education Programs

France appeared to have a much more developed training program for sound engineering than the Netherlands. The respondents with five years of training went through a process equivalent to a master's degree: apart from internships, the program requires the validation of a thesis in the last year of study. Some schools providing long training in sound engineering are specialized in music, theatre, or movies. Furthermore, about thirty public schools deliver a diploma of higher education as technician with an audiovisual specialty. This two-year training is intended to develop the practical skills required to resolve complex technical issues. It is appreciated in the professional field, as well as the respondents who went through it. Private schools, such as the School of Audio Engineering (SAE), a worldwide network of schools based in England and transplanted worldwide, deliver a degree in one, two or three years, but their diplomas are generally less valued in the field than the ones of public education.

In contrast, in the Netherlands, private schools provide the highest possible, if not the only, level of education in this field in the country. Many Dutch respondents were actually trained at the SAE, and explained it was the best possible school in the country. It has been hard to find an actual public training in the profession, and only a couple of respondents had ever heard of any public education program. One was quoted above and gained a professional diploma in general media training in three and a half years mainly composed of internships. The other one is held by fine-art schools and is four years long. One interviewee mentioned being denied admission, but ironically started to work as an apprentice in a studio after this…with the exact same teachers he would have had in the school.

The French informants have thus been through more formal training than the Dutch ones (see Table 1 in "Introduction" chapter). Concomitantly, French respondents tend to more easily switch positions: while the average number of type of positions requiring different skills occupied is 4.4 in the French respondents, the Dutch respondents amounts to 3.7. Although this result cannot be generalized, it is a sign that French respondents of the panel can adapt to a higher variety of jobs, which is useful in the context of a project-organized labour market. It is tempting to explain this difference by the difference in formal education, but this finding would need be formally confirmed by a further study. However, it is possible to suggest that advanced formal education increases the transferability of skills from one art world to another, by granting general knowledge facilitating a technician's adaptation in a new context.

Distribution of "Artistic" and "Technical" Tasks

The comparative analysis of national subsets showed a more developed division of work in France. First, the valorisation, in recent Dutch cultural policies, of "cultural entrepreneurship" is visible in the truth tables. If we compare the truth tables of the outcome "autonomous aesthetic decisions" for French and Dutch subsets, we can see that Dutch management positions are strictly associated with this outcome, whereas it is not the case in the French panel (Table 3.6).

Dutch respondents need to employ their entrepreneurial and bureaucratic skills in order to offer creative services, whereas French respondents are used to a salaried position. Logically, management tasks are realized in parallel with creative tasks in the Netherlands, whereas in France, management tasks are handled by administrative personnel.

We can see that consistency scores for contexts, positions and genres on the outcome "presence of aesthetic decisions" are systematically higher in the Dutch panel than in the French one (Truth tables in Appendix E.1). This means Dutch respondents tend to be more easily granted aesthetic responsibilities. Indeed, many

Table 3.6 Consistency scores of positions on the outcome "presence of autonomous aesthetic decisions", national sets

	French consistency scores	French number of cases	Dutch CS	Dutch NC
Aide	0.076	66	0.2	25
Backline	0	2	x	0
Creation	1	12	0.5	2
Image for sound or broadcast	0.500	8	0.714	7
Management	0.286	7	1	6
Musician	1	6	1	2
PA	0.652	66	0.792	53
Studio	0.897	29	1	5
System	0	5	0.5	4
Teacher	0	4	0	1
Technical but not sound	0.250	4	1	1

Dutch respondents mentioned being placed behind a mixing desk after very little training time:

> *"I learned about a studio in [city] called [studio name]. It's [musician], he's a trumpet player, he was pretty big artist here in Holland. He was building his own studio. And via via I heard about him and I said "that's what I wanna go do". So I went there next day and quit the job I was doing and I started help them build that studio. [...] I had a mentor there, of course there was other engineers working there who sort of taught me but...yeah, I was really thrown behind the board. Really quickly (NI, Interview)*
>
> *Z: How did you had the chance to work in [venue]?*
>
> *I just went there and I said "I wanna learn how to do sound. And they said "ok, have fun tonight". And I was there with stuff and I had no idea.*
>
> *Z: But you were not driven by someone? At the place?*
>
> *No, well...there were some really cool people there, and some people that knew how the stuff worked but...in the end, you have to find everything out for yourself." (NC, Interview)*

In line with findings of Chapter 2, we can interpret that Dutch respondents are placed in positions of aesthetic responsibility sooner because there are fewer aides available, due to a higher difficulty in funding such positions in a context where the income gap, due to Baumol's law, is less addressed by national policies, as we saw in Chapter 2. The cost disease acts in a context where fewer compensation mechanisms exist, it is harder to recruit technicians in the Netherlands despite the more flexible employment laws, and incidentally to develop a vertical division of tasks.

This is confirmed by the comparative sufficiency analysis of the outcome "technical role in an artistic context. Indeed, it shows a much more branching division of work in France than in the Netherlands. The difference between the truth tables for position and context is striking (Truth tables in Appendix E.2.1 and E.2.2): the consistency scores are systematically much higher in the French panels than in the Dutch ones, showing that French cases are systematically more associated with the outcome "technical role in an artistic context", regardless of context, genre or position. While certain positions and contexts are systematically associated with technical work in artistic contexts in the French subset, the number of strict non-association of certain positions with technical work is striking in the Dutch one. It is a sign that predominantly technical jobs, where little artistic input is present, are scarcer in the Dutch cases. Dutch technicians are put in charge of aesthetic matters more readily than in France, where it is easier to hire a workforce to perform delegated tasks.

In sum, we can clearly perceive a strong difference in the division of work between the two countries. In France, many more jobs are technical, integrated in a long chain of delegation, whereas in the Netherlands this chain tends to be shorter, and people tend to be placed more easily in a situation where they have to handle aesthetic matters. Furthermore, the perception of what is artistic or technical appears to depend to some extent on the symbolic hierarchies in the local field, as the national perceptions of working in a club showed.

Discussion

Despite being embedded in project-based labour markets, sound technicians' careers appear to be structured slightly differently to those of artists. Artists tend to perform various activities, more or less directly fitting with their artistic aspirations, always in the discipline in which they are specialists (Menger, 1999; Perrenoud & Bataille, 2017). Their evolution tends to be strictly "vertical": they can gain reputation and move up the social ladder of their art world, or they can struggle, stay stuck, and potentially abandon their discipline for a less risky, but also potentially less fulfilling career (Reilly, 2017). Sound technicians also have this vertical mobility within art worlds, but unlike artists, they have the possibility to work in different art worlds, as some of their skills are transferable and can be used in several artistic contexts. The presence of a formal education can also amplify those skills, and thus facilitate sound technicians' mobility.

As a result, sound technicians, and more generally technical intermediaries using transferable skills to work in different art worlds, are confronted with a broad scope of various practices and conventions. This scope goes beyond the one of artists, who cannot travel as easily between art worlds as the technicians. For instance, I noticed during the fieldwork, that it was normal for the same sound engineer to work on an electro concert, a pop concert, and a death metal concert in the same week. I have cited examples of people who work both in television and on music tours. If, following Uzzi and Spiro (Uzzi & Spiro, 2005), we place this finding under a network analysis perspective, it means that technicians create a significant number of ties between art worlds, and these ties are practically unaddressed by the literature. Technicians bridge cultural and artistic practices from art worlds that, taken from the point of view of audiences, artists, and potentially cultural intermediaries, are completely remote from each other. Hence, they have to be a discrete but important vector of the clash of ideas that fuels artistic innovations. It is not possible to address this issue with the data gathered for this study. However, doing so would almost certainly constitute a way to better understand how artistic practices circulate, are confronted, and mixed, in order to finally give shape to innovative artistic forms.

It has been possible to roughly distinguish two kinds of technicians in music worlds, one of which being directly involved in artistic matters. Firstly, those who, despite the fact that they indeed bring to life a performance's script in a material environment, also have to take autonomous aesthetic decisions. In other words, these technicians are taking part not only in the *mise-en-scène* of the performance, but also in the writing of the script, in close collaboration with the artists. Secondly, the other type do not have such responsibilities, but they have to ensure the fluidity of the different exchanges between various actors taking part in artistic matters during the *mise-en-scène* and the performance, and thus that performance sticks to the rhythm that is assigned by the script which is designed to lead to emotional entrainment. They carry the relational work necessary to keep the performance's pace and synchrony, which is best executed when the person is familiar with the art world's conventions, the expectations of different actors present, and can creatively respond to unexpected and urgent problems.

The latter kind of technicians, executing tasks that could qualify as "technical" if we accept that such a term refers to the absence of direct intervention on the performance's script, was much more present in France than in the Netherlands. The French institutional context, with its higher public funding, its de-commodified labour market, and its system of indirect funding of personnel by high-productivity sectors, allows for the recruitment of larger teams. Besides smoothing the *mise-en-scène* of performances, such positions are a way for beginners to step into different art worlds, to test whether they feel artistically involved in them and learn through the mentoring of senior technicians. It is thus likely to support the development of transferable skills, as well as the circulation of practices between different art worlds. Furthermore, many French respondents mention the fact that they use the days on which they are officially not working, but that are compensated by the unemployment insurance, to work on nascent projects which have little or no budget. Conversely, Dutch respondents are more readily placed in position of responsibility regarding the performance's sound, and less "technical" positions are available. However, following the stimulus to engage in "cultural entrepreneurship", they tend to engage more in managerial activities and to extend the scope of their activities through entrepreneurial practices. Nonetheless, material resources, whether financial or manifested as time they have to invest in these activities, are generally mentioned as too scarce in order to fully engage in such activities.

Conclusion

When hired on musical performances, sound technicians are integrated into project teams in which each actor takes a role. They always have to shape the material object of the performance by handling relational and non-relational matters. However, they also either take part in the writing of the script, or just contribute to give it a material form during the *mise-en-scène*. Their professional trajectories are made up of various projects between different art worlds and at different level of responsibility, and they rely on transferable skills to be employed in very different artistic contexts. The distribution of positions where technicians are directly involved in writing the script of the performance, or not, varies significantly as a function of the institutional context in which technicians are embedded. Such variations are likely to influence artistic innovation in ways that need to be further explored.

Relational work, as seen in Chapter 1, is important for their tasks. A certain degree of closeness is necessary to ease creative artistic collaboration, and smoothing relations is the very core of more "technical" positions, in which technical intermediaries do not take part in the script's writing. However, in order to be performed, this work needs to be framed by a set of conventions guiding the expectations of each actor to the performance, that can serve as standard reference point and help them to do their job properly. This is what we will see in Chapter 4.

The Power of Sound Technicians over the Script: Negotiations around the Shape of the Performance's Object

4

The previous chapters, especially Chapters 1 and 3, have shown the essential aspects of the work of sound engineers: they materialize a script designed by artists and cultural intermediaries in a physical space in order for it to reach an audience. In doing so, they might take aesthetic decisions. However, as they shape the object that will be the target of the attention of all participants of the performance, their actions are in turn scrutinized by those who have written the script, as well as the audience. The expectations of artists, audiences, and cultural intermediaries on the object of the performance can align harmoniously, facilitating the work of sound engineers. However, whenever this is not the case, the latter have to produce a sound that reflects a compromise between the demands of artists, cultural intermediaries and audiences.

This compromise, therefore, reflects the expectations resulting from the script, i.e. "choices about the paths [actors] want to take and the meaning they want to project (Alexander, 2004, p. 550). The script, as a sort of scenario, is collected from the background culture, and the performance finds its meaning through this influence. Hence, the object produced during the *mise-en-scène* must be faithful to the script. However, actors can have diverging views on what the performance should be. As a consequence, sound engineers can be faced with contradictory demands, for instance when the artists want their music to be heard loud and clear, and the organizer of the concert is obliged to obey a legal acoustic pressure limit. The decisions that technicians make will thus reflect the power relationships in the workplace, notably the power to decide what the performance's script is. The

Electronic supplementary material The online version of this chapter (https://doi.org/10.1007/978-3-658-33029-3_4) contains supplementary material, which is available to authorized users.

69
A. Battentier, *A Sociology of Sound Technicians*, Musik und Gesellschaft, https://doi.org/10.1007/978-3-658-33029-3_4

view of one specific actor can prevail over another, or technicians can potentially manage these choices themselves.

As technicians practically shape the performance's object, observing how the work of technicians is driven during the *mise-en-scène* is a way to understand how power over the performance's meaning is distributed among participants. Artists, cultural intermediaries and audiences will express directly, or indirectly through various reactions and behaviours, whether the performance's object corresponds with their idea of the script. They will accordingly require that the technicians make changes in order to make the performance's object fit this idea. Technicians can grant these demands, diligently or reluctantly, discretely ignore them, or reject them more or less vehemently. They can also interpret the different reactions of others in order to address their expectations. However, they will, at some point, necessarily shape the object to match their own idea of the script: they are, after all, the ones using a remote control which nobody else is able to use. This aspect of their work is the root of their autonomy and unique power in art worlds. They can be pressured to make the performance's object correspond to a particular view of the script. However, their contribution is needed, as they have to engage their competencies in order to achieve any required result.

In this chapter, I will argue that the distribution of power over the performance's meaning is related to the music genre characterizing it. By "music genre", I do not mean jazz, classical music, rock, pop, electro or reggae, but I refer to the sociocultural notion of genre developed by Jennifer Lena (2012), which classifies music as "scene-based", "industry-based", "traditionalist", or "avant-garde".

Theory: Music Genres of Performances

According to Alexander (2004), individuals in society evolve in a conscious and unconscious magma of systems of collective representations that form the deep background of social life. People intending to set up a performance select consciously and unconsciously a subset of elements from this general background in order to imbue the performance with an intended meaning. Alexander calls the resulting selection a script, an "action-oriented subset of background understandings" (Alexander, 2004, p. 550). The script, embodied in a physical object (the performance) produced during the *mise-en-scène*, can be sketched out purposefully before the performance, for instance in the case of a theater play, or can be reconstructed retrospectively, for instance in the case of a social drama embedded in daily life. In musical performances, the script's embodiment is handled practically by technical intermediaries during the *mise-en-scène* phase, a moment of staging consisting of the "bringing together or confrontation, in a given space and time, of different signifying systems, for an audience" (Pavis, 1988, p. 87; cited by Alexander, 2004, p. 554). This moment, in a live music performance, starts when technical intermediaries begin to build the stage. However, they are not deciding purely by themselves how the actual show will be delivered. Indeed, artists and cultural intermediaries generally jointly conceive most of the script, and sometimes all of it. They have thus built expectations upon what the performance shall resemble, and will express them to technicians. Audiences also have expectations, and they will express their satisfaction or disappointment through a range of reactions such as shouting, clapping, boos or whistling. They can also come to the mixing desk during the concert and directly ask the sound engineer to change the sound.

How do sound engineers arbitrate between all these pressures? When there are contradictory expectations, who gets priority? Can sound engineers impose their decisions against the wishes of another participant? Becker argued that, in an art world, although an artist is "the person who performs the core activity without which the work would not be art" (Becker, 1982, p. 24), "aesthetic conflicts between support personnel and the artist can also occur" (Becker, 1982, p. 25). He takes the example of motion pictures, the production of which implies "multiple difficulties of this kind: actors who will only be photographed in flattering ways, writers who do not want a word changed, cameramen who will not use unfamiliar processes" (Becker, 1982, p. 26). I discussed in the "Introduction" chapter the conceptual difficulties posed by the fuzziness of the notion of support personnel. However, Becker points that authorship of an artwork, in the sense of being identified as the artist producing it, does not grant full control over its realization and by extension, the meaning that it is intended to carry. This meaning results in fact from the actions of all actors in the art world, including "support personnel" or, as I prefer to call them, technical intermediaries.

Ethnographies of music worlds display a diversity of situations in which musicians are more or less in control of the shape of the final output of their work. For instance, Rudent (2008) shows how the power to decide the shape of music was distributed between a sound engineer, the label, and musicians during the recording of the first album of the French singer *Mademoiselle K*. Perrenoud (2007) mentions the pressure put on one of his bands by hasty recording technicians, unhappy about the time taken to record the album and who do not want to work overtime. Kealy (1979) discussed the evolution of music recording in the US. He distinguished three modes of collaboration between artists and technicians. In the craft mode, musicians play and are forbidden to do anything else, the rest being the realm of technicians. Conversely, in the art mode, musicians decide about everything, and can even use recording techniques themselves as a creative tool. Finally, the entrepreneur mode is a middle ground between the two, in which musicians and technicians engage in an equal partnership with shared control over the output.

How is the power to decide the final shape of a music object distributed? Part of the answer could be that music genres are traditionally understood to be the classification of musical objects according to their musicological properties, such as tones, types of instruments used, and specific rhythms. Music genres are symbolic boundaries, "conceptual distinctions made by social actors to categorize objects, people, practices" (Lamont & Molnár, 2002, p. 168). They are omnipresent in music and used for various purposes. For instance, cultural intermediaries emphasize the necessity to associate a musical production with a genre (Jeanpierre & Roueff, 2014; Lizé, 2016a; Maguire & Matthews, 2012; Negus, 2002), in order to successfully promote it towards potential audiences (Hitters & van de Kamp, 2010; McLeod, 2001). Music genres are also used by critics (Schmutz, 2009; van Venrooij, 2009). Finally, they are of crucial importance for musicians, who use them to communicate and identify the conventions on which they will base their collaboration (Faulkner & Becker, 2009).

However, in music worlds, music genres converge only partially with social boundaries, "objectified forms of social differences manifested in unequal access to and unequal distribution of resources […] and social opportunities" (Lamont & Molnár, 2002, p. 168). For instance, Perrenoud (2007) showed that a musician's job portfolio is generally composed of several music genres. Vlegels and Lievens (2017) showed that music genres have a trivial influence on taste stratification in music, related to the effects of age, and of the artists' popularity and international influence. Schmutz (2009) perceived that the evolution of music genre classifications did not impact the unequal gender distribution of media attention in a dozen national newspapers in France. Using data harvested from online platforms, Van Venrooij and Schmutz (2018) showed that classification ambiguity does not impact critical success, and even enhances it if the album comes from the field of restricted ("small-scale") production (Bourdieu, 1971).

Therefore, the notion of music genre shows significant limits when used to explain the social boundaries of music worlds, and thus can hardly be used to explain the distribution of authority over the shape of the object of a musical performance. Lena explains these limits as:

> Cultural sociologists, and the sociologists of music who are my focus here, often treat genres as natural objects for the sake of expediency and to favor the emic experience of fans, artists, and other 'insiders' in music communities. Stories about socio-musical

identities are usually premised on the assumption that patronage of a musicological genre makes one's experiences and opinions coherent. [...] But such 'natural' philosophies of genres don't hold up under scrutiny; they 'collapse a complex, shifting social world full of debate and disagreement into an inevitable chain of events leading to the present, during which necessary transformations take place' (Lena 2012: 146). Claims for inalienable boundaries between styles cannot be sustained in the face of substantial evidence that stylistic boundaries are social constructions accomplished when people and organizations collaborate in order to re-/produce genres. (Lena, 2015: 150)

In order to address these limits, she developed a sociocultural conception of music genres (Lena, 2012). In her book, she analysed the historical evolution of fifty-six genres in the United States during the 20th century. She identified four configurations of twelve dimensions related to socio-economic parameters and the process of symbolic bounding of music styles. These configurations, "avant-garde", "scene-based", "industry-based", and "traditionalist", constitute what she defines as music genres (Table 4.1).

In her theory, the usual use of the notion of "genre" is rendered by the term music "style". A music genre is a "systems of orientations, expectations, and conventions that bind together industry, performers, critics, and fans in making what they identify as a distinctive sort of music" (Lena, 2012, p. 6). This definition covers both the socio-economic conditions influencing musical arts worlds' development, and the specific musical and aesthetic features to which genre *aficionados* are symbolically attached. Lena's notion of genre, however, is constructed on socio-cultural features, such as behaviours and membership dynamics, and is not based on musical content such as instruments used, typical tones or rhythms. As such, it is close to the notion of "script" of a cultural performance, defining the set of elements drawn from a larger cultural framework and assembled to produce a meaningful performance. In this chapter, I will argue that the distribution of authority over the shape of the performance's object between artists, audiences, technical and cultural intermediaries can be explained by Lena's typography of music genres.

Method: Ethnography and QCA

During the *mise-en-scène* phase, technicians act on a series of demands reflecting the expectations of the three other types of actors of what the performance's object will become. It is necessary for the other actors to be satisfied with the object. They can achieve this by addressing these demands directly to the technicians, but they can also simply express their satisfaction or irritation. Technicians as

Table 4.1 Dimensions of music genres (Lena, 2012: 9)

Dimension	Genre forms			
	Avant-Garde	Scene-based	Industry-based	Traditionalist
Organizational form	Creative circle	Local scene	Established field	Clubs, associations
Organizational scale	Local, some Internet	Local, Internet linked	National, worldwide	Local to international
Oraganization locus	Homes, coffee shops, bars, empty spaces	Local, translocal, and virtual scenes	Industrial firms	Festivals, tours, academic settings
Sources of income for artists	Self-contributed, partners, unknowing employers	Scene activities, self-contributed	Sales, licensing, merchandise, endorsements	Self-contributed, heritage grants, festivals
Press coverage	Virtually none	Community press	National press	Genre-based advocacy and critique
Genre ideal or member goals	Create new music	Create community	Produce revenue, intellectual property	Preserve heritage and pass it on
Codification of performance conventions	Experimentation	Codifying technical innovations	Production tools that standardize sound	Hyper: great concern about deviation
Technology	Experimentation	Codifying technical innovations	Production tools that standardize sound	Idealized orthodoxy
Boundary work	Against established music	Against rival musics	Market driven	Against deviants within
Dress, adornment, drugs	Eccentric	Emblematic of genre	Mass-marketed "style"	Stereotypic and muted
Argot	Sporadic	Signals membership	Used to sell products	Stylized
Source of music name	Site or group specific	Scene members, genre-based media	Mass media or industry	Academics, critics

well have expectations regarding the other participants and need to communicate them. Scrutinizing these behaviours during the *mise-en-scène* and the performance provides insights into who is satisfied and who is disappointed with the final shape of the performance. Scrutinizing whether technicians grant, ignore or reject these demands is a way to understand who has the power to decide the performance's meaning.

In order to understand whether Lena's music genres[1] explain the distribution of authority over the performance's object shape, I will perform an analysis of these different reactions, behaviours and arbitrations recorded in the observation notes, and test whether the genres are able to explain their coherence. In order to do so, I assigned a genre to each observation (Table 4.2), using the empirical elements presented in the table of observations in the "Introduction" chapter.

During participant observations, I recorded the different reactions that were visible to the technicians. In the moment of *mise-en-scène,* this perspective is appropriate for understanding whether the shape taken by the performance's object is considered satisfying by the different actors. Indeed, artists and cultural intermediaries directly express their disagreements, their specific needs, and can potentially compliment technicians if they like the shape given to the music. Following Bechky's accounts of the distribution of roles and hierarchies in temporary teams attached to movie productions (Bechky, 2006), I consider: thanking, admonishing and joking as signals of whether a participant is acting accordingly in his/her attributed role. They will communicate their needs and expect the sound to be modified as a consequence. Although in a different manner, audiences also send signals of this kind. Beyond the conventional signs of enjoyment or dislike such as clapping, booing, singing and so on, it is not unusual that audience members come up to the mixing desk and make demands or compliment the sound engineer. Furthermore, artists and cultural intermediaries also rely on conventional signs of enjoyment or dislike in order to express themselves. Some of these expressions can be demands disguised as emotional expressions. My experience as a former technician helped me to recognize such moments and to identify whether the sound engineer positively responded to it or not. Finally, respondents, and potentially artists, cultural intermediaries or audience members, often shared their feelings about the ongoing actions with me, thus providing more insights into whether they found the actual performance satisfied their expectations.

---

[1] In the remainder of this chapter, the term "music genre" will refer to Lena's notion of genre. For the usual understanding of the term, I will use the term "music styles".

Table 4.2 Genres of observations

Ref Obs	N01	N02	N03	N04	N05
Genre (CF Lena)	Scene-based	Industry-based	Industry-based	Scene-based	Industry-based
Ref Obs	N06	N07	N08	N09	N10
Genre (CF Lena)	Avant-Garde	Traditionalist	Scene-based	Scene-based	Avant-Garde
Ref Obs	F01	F02	F03	F04	F05
Genre (CF Lena)	Industry-based	Scene-based	Industry-based	Industry-based	Industry-based
Ref Obs	F06	F07	F08	F09	F10
Genre (CF Lena)	Scene-based	Scene-based	Traditionalist	Traditionalist	Avant-Garde
Ref Obs	F11	F12	F13a	F14	F15
Genre (CF Lena)	Scene-based	Scene-based	Traditionalist	Scene-based	Avant-Garde
Ref Obs	F16	F17	F18	F19	
Genre (CF Lena)	Scene-based	Scene-based	Scene-based	Industry-based	

All these reactions noted during the observations were coded as "positive" or "negative". A positive reaction is interpreted as a correspondence to the script as perceived by the person doing the expression, while a negative reaction is interpreted as a deviance from it. These expressions can be irritation due to a delay, satisfaction produced by the quality of the sound, an ecstatic yell due to a well-managed tricky musical part, public display of complicity bonds, nervousness, a demand formulated aggressively or conversely in way that demonstrates complicity, a strong audience reaction (dancing, shouting etc.), or conversely the display of indifference and/or hostility (weak clapping, booing, disparaging comments, intensive cell phone use etc.). Table 0-28 in Appendix F.1 gives the number of expressions noted on each observation. Furthermore, I took note of how technicians responded when confronted by such demands, i.e. whether they diligently granted the more or less explicit demand, whether they discretely ignored it, or if they more or less curtly rejected it. Finally, following Kealy's model of distribution of authority over the sound of a music piece (Kealy, 1979), I captured whether technicians' decisions were taken autonomously, in negotiation with another participant, or if they were imposed on them.

Once the expressions were coded as positive or negative, I used qualitative comparative analysis (Ragin, 2008; Rihoux & Ragin, 2009; Schneider & Wagemann, 2012) to identify the most significant cases, i.e. those that contained the largest proportions of expressions of one actor in the observation. I calculated each expression type's (who expresses and whether the expression is positive of negative) proportions of expressions in the total of expressions noted in the case. I then converted these proportions into fuzzy sets membership scores in sets of "observations with a high rate" of positive/negative expression of one type of actor[2]. For example, in F11, 40% of the total of noted expressions were positive and coming from technical intermediaries. F11 has the highest membership score in the set "observations with a high rate of technical intermediaries' positive expressions". It means that in this particular observation, technical intermediaries' positive comments represent 40% of all noted expressions, and that this proportion is the highest observed in all observations.

---

[2]These scores were obtained following the direct method (Ragin, 2008), with the median as a crossover point. The threshold of full membership (corresponding to a minimum score of 0.95), was empirically defined, as a gap between the small number of high proportions and the ones directly below them generally emerged. The full non-membership score was defined just below the median. Thus, the membership score quickly decreases for values below the median, which allows us to focus on the observations which are richer in noted expressions. The membership scores are summarized in Table 0-29 in Appendix F.2.

   In doing so, I sorted the cases by the ones in which the interpretation of the quantity and content of expressions was the least ambiguous. A content analysis was then performed on the cases in which the membership score was above 0.8, i.e. where the volume of expressions of each type of actor regarding the others was the most important (Table 4.3). The contents of expressions noted during these observations was analysed, with the intention of understanding if they showed satisfied or disappointed expectations of each type of actor, and by extension how they were the signs of a script reconstructed through these manifestations of mood. In a nutshell, qualitative comparative analysis identifies the cases where the most meaningful material is available, and content analysis shows where there is a high volume of meaningful material in these cases. Used in combination, these data allow one to access the script of the performance, and to understand whose perception of the script is given priority when it comes to giving the performance's object its meaningful shape, as well as the mechanisms of attendant negotiations.

Table 4.3 Observations with a membership score over 0.8 - OWHR = Observations with high rate

	Scene-based	Industry-based	Traditionalist	Avant-garde
OWHR TI POS	F11, N01, F14, F02	F05, F03	F08, F09	N10
OWHR CI POS	N09, N04, F11, N01, F18	F04	N07	None
OWHR ART POS	F12, N01, F07, F14	F04	F13a, F09	N10, F15, F10
OWHR AUD POS	F17, F14	F03, N03	F13a	None
OWHR TI NEG	F02, F06	F19, N02, F05, N05	F08	None
OWHR CI NEG	N08, N04, F06, F16, F17, F18	F03	N07	None
OWHR ART NEG	F12, F07, F02, F17, N04	F01	N07	F10
OWHR AUD NEG	F14, F06, F18	N02, N05, F04	N07, F08	None

Distribution of Authority over the Performance's Object Shape do Follow the Genre Boundaries Drawn by Lena

Lena's genres efficiently explain how the final responsibility over technical matters in musical performances is distributed. First, cases of the same genre empirically display the genre dimensions described in Table 4.1. The genre appears therefore to accurately describe key elements of the script of a music performance. Second, the distribution of decision power over the shape of the performance's object also follows the dimensions of music genres. In this section, I will show how participants' behaviours, expectations, and power over the performance's meaning can be understood through the prism of each genre's dimensions.

Scene-Based Genre

According to Lena, actors of this genre intend to build a community. Its conventions are aesthetically well established, although not definitively fixed. Listeners are sensitive to the codification of the style, i.e. the manner in which the conventions of the musical style are practically implemented. In line with this definition, I have noted in this genre a strong tendency for the cultural and technical intermediaries to express their opinion on musical quality, whether in positive or negative terms. This tendency to share their opinion can be influenced by the fact that their financial compensation is generally low, and potentially non-existent. This compensation becomes then symbolic, derived by being more involved in the aesthetic definition of the performance.

In this context, many conflicts between artists and technicians about the sound were present in these kinds of observations. The latter can potentially try to impose their aesthetic preferences on the opinion of artists, and sometimes manage to do so. Another sign of this community-based dimension of scene-based genres is the pressure exerted on musicians, particularly with respect to the schedule. Rules are defined according to the needs of the community; these generally prevail over the specific needs of artists. Eccentric demands or schedule modifications are generally poorly received and rarely granted. For instance, in F06, the unannounced delay of one of the bands outraged one of the venue's technicians, who attacked the concert organizer verbally, and personally 'sermonized' the musicians when they finally arrived.

The welcoming technician for the light, which is also the technical chief[3] of the place and stage manager*[4] for this night, is furious about the perspective that his break won't be at 7PM sharp. They argue with the organizer because the first band to play, a choral is not here on time, just like last year. […]

7.10 PM, end of the soundcheck. "Shall we leave it like this? -YEAH !!!!" answers the technician who is really, really angry. "Shall we install the choral when they arrive?—No, they really have no respect." […]

The choral arrives at the time of the show (8.30). They'll play second, and they'll have no soundcheck. The technician is having an explanation with the band. They are visibly arguing but less strongly than earlier. (F06, fieldwork notes)

However, following the schedule is not a constant in this genre's cases. Indeed, many of them (N01, F07, F11, F14), happened in a relaxed atmosphere despite significant delays, that did not seem to bother anyone. In sum, the "community", i.e. the group of people present in the concert venue, and its internal rules, decides whether a delay is a problem or not, but this decision is never monopolised by a particular type of actor. Another sign of this community-based dimension of the genre is the fact that artists, technicians and cultural intermediaries generally eat together before the concerts.

The focus on codification of the musical style, and participative dimension, is also present in the attitude of the audience. People often comment on the quality of the sound, sometimes by directly addressing the technician. Appreciation is expressed by a moderate enthusiasm and by focused listening: conversations can be intrusive and be actively discouraged by other members of the audience, as for instance in F14. In this observation, a jazz event in a jazz club in Paris, there was a bar at the rear of the room, where a couple of people were having a chat. A person came out from the audience and told them to stop talking. This person was visibly approved of by the surrounding people, and the two chatterboxes stopped talking. Musical dislike can be passionately expressed. For instance, during F18, a two-day festival* mixing classical and pop music in a church, a two-minute delay during the soundcheck of an orchestra of *chanson française* provoked an outcry from the audience waiting for the classical organ performance programmed after it. In this case, the reaction is a sign of the community defending against the intrusion of a rival genre.

---

[3] The technical chief (*directeur technique*) is the person who oversees the technical teams and gear of a specific place or event, as defined by the board of the venue and the artists.
[4] The stage manager is there to ensure that everything that is needed on stage is provided, to answer unexpected needs of artists, and that the day's schedule is respected.

The soundcheck is running late. The venue manager comes and says that it must be over quickly, as the organ player of the church is planned at 5PM. [...]

5.01 PM, the audience starts to pressure to stop the soundcheck. One of the technicians gets angry: "Quiet les fillonistes[5]! We just have one minute of delay! Go get refunded!". The concert has no entrance fee. [...]

After a few minutes, the audience gets louder. Someone yells "We want the organ!". Someone behind the mixing desk says that it is not normal that such a band plays here, in a way intending for the front of house technician to hear her. (F18, Fieldwork notes)

Here, the organist's audience is getting very clearly and quite aggressively involved in shaping the performance; the presence of pop music is merely tolerated by them. They perceive the schedule disruption as an abuse, and feel authorized to express themselves to pressure musicians and technicians to respect the schedule. In this particular observation, dimensions of traditionalist genre associated with the classical organ performance, are intertwined with the ones of the scene-based characterizing the whole festival*. Hence, the participatory tendency of audience of scene-based genres is exacerbated by the idealized orthodoxy of traditionalist dimensions coupled to the organ performance. However, in general, in scene-based observations, audiences tend to try to influence the performance's script, much more than in the other genres.

In this genre, the power over the shape of the performance's object is distributed among all participants. Technicians can impose their views, artists are limited in what they are permitted, and audiences can pressure these actors in order to obtain a modification of the show.

Industry-Based Genre

Lena writes that an orchestra producing music belonging to the industry-based genre intends to "produce revenue by selling musical products to as many consumers as possible". This music is "highly codified, driven by industry categories and the production tools that standardize sound" (Lena, 2012, p. 41).

As a consequence of this strong codification, professionalism appeared to be a dominant characteristic of participants' exchanges. Division of work, whether horizontal or vertical, is strict and explicit. An aide will not discuss the decisions of the chief, and will take care to always be (or at least, look) busy. On the

---

[5]Supporter of François Fillon, the right-wing presidential candidate, close to the Catholic community, in the midst of a political scandal at the moment of the observation.

horizontal side, the sound engineer will restrain him/her-self to comment on an element which does not *a priori* belong to his/her domain of competence, such as an out of tune guitar (N02, F04). Resentment, and potentially rejection, can be triggered by the trespassing of professional boundaries. For instance, in F01, the studio sound engineer of the band came to the mixing desk several times to give advice to the front of house* sound engineer. The latter, despite never having worked with this artist beforehand, clearly perceived these proposals as intrusive. Upset, he coldly suggested the former leave the mixing console to him. As the studio engineer did not knew how to use it, and as his role was not to handle the front of house* sound, he quickly stopped providing advice. The other technicians, witnessing the scene, considered that the behaviour of the front of house* engineer was legitimate.

Unlike scene-based observations, comments on the quality of the music stay private, and are potentially only expressed to close colleagues. This criticism, that can nonetheless be very intense, focusing generally on respecting the role expectations rather than on music aesthetics. Cultural intermediaries often mediate between the band and the venue's team and thus must avoid disparaging comments. Technical intermediaries are inclined to say that their personal opinion does not matter, and that their role is to produce the best possible sound regardless of their opinion on the music. However, the definition of "best possible sound" is subjective and varies with different bands, places, and people.

In this same logic of strong division of work, technicians expect artists to clearly define their needs and aesthetics expectations in terms of front of house* and monitor* sound. A lot of technicians' negative comments are complaints about poor communication of the artists' needs. However, the artists are less required to restrain their individual and potentially eccentric needs. In F05, the band just came on the stage and rehearsed during the soundcheck, the engineers had no possibility to interrupt them to ask for a specific instrument. Basins of pig's blood were available for the ritualistic shower of the members of a metal band in F03[6], and a stagehand* had to run out before N03 to buy balloons and streamers for the birthday of a band's member.

The intern goes out for a smoke. Before he leaves, the venue sound technician tells him to find a vacuum cleaner to clean the amps grids. Meanwhile, he tells me he needs to clean the pig's blood from the monitor*. While rubbing the speaker, he explains, amused by my intrigued face: "It's a metal band that came yesterday, they put a mess,

---

[6]The demand came from the band of the day before, that can also be categorized as industry-based. It was mentioned during the observation by the venue's technician who was cleaning some blood from the monitors* that had been forgotten by the dedicated tour cleaning team.

it was a massacre. There were dead animals on stage, it was smelling and all...the guys poured basins of pork blood on their head before the concert, you have showers in the backstage. But they have a whole team with them, to clean afterwards. They did it well, but they forgot my monitor*. " (F03, Fieldwork notes)

During the dinner break, the stagehand* tells that he was about to take his train back home when he was called to get inflatable balloons for the birthday of one of the headline band members. (N03, fieldwork notes)

Audience's physical reactions are much stronger than in the scene-based genres when the music is appreciated. During F03, the 700 people of the audience were dancing so intensely that drops of condensed sweat were dripping from the roof. In N03, teenagers screaming as the singer appeared effectively produced a sudden gust of wind. Applause saluting the end of a tune were also quite intense. When the audience, conversely, was not caught up in enthusiasm, numerous conversations occurred here and there. These conversations were generally well tolerated by the other audience members, even if their volume competed with that of the music itself. Contrary to the scene-based genre, audience members seemed generally less focused on the musical aesthetics, and more on sharing and experience with their peers.

Industry-based genre observations are thus much less participative than their scene-based counterparts. The scripts are defined by artists and cultural intermediaries, technicians implement them to the best of their abilities and must avoid modifying them. Every actor has a strongly defined role, must fulfil the expectations attached to it and do not exceed them. Audiences' influence is reduced to the presence or absence of a strong emotional involvement.

Traditionalist Genre

Participants in traditionalist genres mainly try to preserve and transmit a heritage and are concerned about internal deviancy (Lena, 2012, p. 9). "A preservationist spirit is precisely what strongly differentiates traditionalist genres from other genre types" (Lena, 2012, p. 48). Traditionalist music is associated with styles that have declined after a period of success, maintained by groups of passionate people, brought together by the nostalgia for an idealized, and potentially fantasized, golden age and the celebration of past idols. Lena excludes classical music from this genre, considering that this category only includes styles that have had a period of commercial and industrial success (Lena, 2012, p. 47). Classical music, in this regard, is indeed different in nature. However, some forms of listening to classical music, implying respectfully silent audiences focusing on faithfulness to the score of a piece, although the instruments' evolution, venues' acoustics, musical training and modern hearing sensibilities prevent the possibility of an "authentic" historical reproduction. It is hard, therefore, to not apply the characteristics of traditionalist genre to classical music: preservation and continuation of heritage, concern about internal deviation, importance of scholars and critics, an idealized past etc.

The work of technical intermediaries in classical music must be as transparent as possible. In my only observation of a concert of this style (F08), microphones were specifically selected for their visual discreteness. Artists were free to redefine the terms of the performance at any moment: when the conductor decided at the last minute to change the orchestra's placement, he did not take into account that technicians had barely 15 minutes to unplug, move, and re-plug their gear before the beginning of the rehearsal, during which no member of the technical team was tolerated on stage. Moreover, the chief sound technician had to run after the conductor to hear, just in time, where the different sections were to be placed, as he was addressing the stage manager* during the process, and paid no attention to the sound technician.

> We are back to the stage, there are about 10 people here. The chief recorder salutes the conductor. He decides to change the orchestra disposition. Immediately, the stagehands* move the chairs, and the microphones must follow as they can. While the conductor describes the changes to the stage manager*, the chief recorder must run behind to be informed of the changes. The assistant says "well, little twist, otherwise we could have taken a coffee."
>
> 3.35 PM, they must unplug and re-plug everything. An old man, well dressed, arrives, I think it is the choir conductor. Choristers starts to arrive, one by one.
>
> 3.48, the flutist arrives. The chief conductor and her two assistants are still repositioning microphones. The stage is getting busier. Violins are tuning, the choir is installed. There is a sort of rumble, everyone is setting up. The chief recorder and the assistants are running everywhere to install their microphones.
>
> 4.03. The rehearsal starts, we are back in the studio. Many microphones have been hit or moved when the musicians got installed. (F08, Fieldwork notes)

Although only one observation was made in classical music, several interviewees worked at least partially in it. All of them mentioned the importance to "sound acoustic", meaning that their intervention must give the illusion of transparency

> "[Classical musicians] have the feeling that when you will amplify their music you will impair the sound. So from the beginning you're an intruder to them. […] For the

classical musician, you need to be discrete, to not be seen, to not be heard...people have to believe it's natural." (FQ, interview)

"When you amplify a classical orchestra and that it has to sound like we were in an auditorium...you, that's your bill of specifications, it has to sound acoustic." (FN, interview)

One of the respondents suggested that this rejection of amplification is based on the fact that classical music predates the invention of sound reproduction technology:

"For classical, at first, sound taking is a heresy. Because it's a modification of what they do. Because back in time...classical music has always been rendered acoustically." (FH, interview)

This approach to sound corresponds to the "idealized orthodoxy" suggested by Lena for the relationship with technology in traditionalist genres (Lena, 2012, p. 9), and reinforces the membership of classical music to it.

Three other observations of traditionalist genres have been made. F13a is the concert of a former star of *variété française* from the 1980s, which attracted a few dozen fans in a little Parisian theatre. He was playing his hits, with the instrumental part burned to a CD which was triggered by the venue's technician by the phrase "*Maestro, quand tu veux*" ("Maestro, your call"), typical of the practices of this style and era. N07 was an open mic of covers of hits from recent decades: one of the musicians mentioned that this music was the one that his father, and not he himself, listened to. Finally, F09 was a tribute to a poet, whose texts were sung or recited in a series of *tour de chant*.

Despite the apparent differences between these cases, I perceived the characteristics of traditionalist genre. Concern about deviation was clear from the strongly polarised comments on musical quality, either ecstatic (F08, F09) or very critical (N07). The critique also included sound quality. Furthermore, these observations were the only ones where my presence as an observer was sometimes poorly received. In F09, the technical manager of the theatre wanted first to drive me out, but finally accepted my presence on the condition that I stay in a corner and did not touch anything. In N07, one of the organizers of the event perceived me as a freeloader, reminding me twice that I could donate to the association organizing the event.

Artists are treated in a similar way to the ones from industry-based genre. A certain eccentricity is allowed, and they are treated with a form of deference, potentially related to their status as conduits of heritage. I have not witnessed a direct critique from the technicians concerning respecting professional boundaries, even in private exchanges and when the artists have added a significant additional workload (F08). In F09, despite the fact the day's schedule was clearly overloaded for the technician in charge of all of the system installation, stage managing, front of house* and monitor* mixing, he stayed calm and was not excessively scathing in comments addressed to the musicians and his colleagues; while the musicians spent a significant amount of time in mundane conversation during the extended soundcheck time given.

The audience stayed static and attentive as they listened. People expressed their enthusiasm by affable and non-excessive applause. If they did not like the music, it was not explicitly expressed. In F08, some people in the first row, which was visible from where I was watching the concert, clearly looked bored. One of the spectators even slept for a third of the performance. However, everyone stayed still, and applauded at the end. During N07 amateur orchestras often had a variable mastery of their instrumental and singing parts, and the results were sometimes hard to bear. Nonetheless, the audience applauded at the end of every tune and stayed in the room. However, people were clearly disengaged from the show and had loud conversations. Unlike the hostile reactions that have been observed in scene-based genres, people do not directly attack the artists by booing, or shouting negative comments. They withdraw from the performance, but do not try to stop or undermine it.

Artists are central in traditionalist genre. They are almost sanctified and can change the script of the performance almost as they please. Technicians are required to follow these changes, almost irrespective of the consequence they have in term of workload. Audiences also perceive the artist through this form of sanctification and maintain respectfully focused regardless of the degree to which they actually like the performance. However, criticism can be harsh, as the sanctification of artists comes with high expectations of their performance.

Avant-garde

In Table 4.3, we have seen that very few "avant-garde" cases have a membership score over 0.8 in all types of sets, except the expressions of artists. Table 4.4 contains the average number of noted expressions by type of actor and genre, and shows that except for artists, actors participating in "avant-garde" performances express themselves less than in other genres.

Table 4.4 Average number of noted expressions by observation, sorted by genre

	Obs Avantg	Obs. Indusb	Obs. Sceneb	Obs. Trad
Negative expression ART	2.75	0.625	2.00	0.5
Negative expression AUD	0	1.125	0.77	0.5
Negative expression CI	0.5	0.375	1.85	1
Negative expression TI	3.25	9.625	6.38	5.75
Positive expression ART	2.25	1	1.62	2.25
Positive expression AUD	0.25	2.25	1.15	0.5
Positive expression CI	0	0.5	1.31	0.25
Positive expression TI	0.75	1.75	2.08	2.5
Total	9	16.125	15.92	12.75

In this case, a focus on how technical decisions are taken is necessary to understand the influence of the genre's dimensions. Avant-garde cases are characterized by a near-total absence of negotiated decisions: technical decisions in avant-garde observations are either imposed, or autonomous. This result is in line with the objectives and the level of codification of the genre as described by Lena. People involved in this genre intend to "create a new music" in an environment where the level of codification is weak. The artistic process being experimental, there is no agreed process to which one can refer to arbitrate their choices. Hence, technicians are either strictly directed by artists, or are completely independent. In the latter case, they must find solutions themselves to problems that the *mise-en-scène* process brings. In a context where working routines and conventions are not explicitly established, all participants take risks, as they must find solutions to novel problems. Failure is more likely to happen, but success is more gratifying than in more codified contexts, because it relies more on individual initiative than in tried and tested models of decision-taking. However, the pressure on each actor is more important in the non-avant-garde. It is even more true for artists, who initiate the projects in observed cases, and are perceived as responsible for the quality of the final result. Hence, they express their satisfaction or disappointment in a more frank and prescriptive way than in other genres.

F10 is the last performance of the tour of a contemporary opera of serial music mixing actors' theatrical performance and chamber music. The orchestra and actors were amplified. The conductor expressed his great disappointment at the end of the full-dress rehearsal, including to the sound engineer: he thought that the actors were too loud and overshadowed the music. The fact that the technician had no opportunity to do a soundcheck beforehand was not seen as an acceptable excuse. The significant tension during the general rehearsal did not prevent the show from being "maybe the best we've done so far" according to the director. In these observations, cultural intermediaries were practically invisible.

These relations between producers do not explain the lack of expressivity of audiences. A working hypothesis has been suggested by a stagehand* at F10, who was used to working with this *avant-garde* orchestra which constituted of volunteers and only worked on experimental projects. During a conversation at the dinner preceding the performance, he told me about his experiences with the orchestra. I suggested that serial music might be difficult to appreciate for an untrained ear. He answered that on the contrary, according to his experience, children were generally the most receptive audience for this music style. To him, adults are inhibited by their expectations of what music is, should be, is not and should not be, thus creating a cognitive filter biasing the interpretation of new sounds as music, preventing the appreciation that comes from this alternative interpretation.

Any form of sentimentalism aside; this proposition is interesting as it suggests that avant-garde audiences are looking for a form of deconstruction of pre-existing expectations. The point of going to avant-garde performances, interestingly, is precisely to have a surprise, a diversion from normality. Hence, challenging existing expectations. Such deconstruction, if successful, can generate a space for the reconstruction of a scene-based genre, which is a potential evolution of avant-garde performances, according to Lena. But during the performance, it can leave the spectators quite puzzled, which can explain the relative lack of expressivity of audiences in these performances.

The uncertainties of avant-garde performances imply that participants, and especially artists and technicians who jointly work to give a material shape to music, share the related decision-making burden. A final arbitration is done by the people endorsing authorship. This arbitration is of binary nature: it is either good or bad, but the way to obtain one or other judgement is uncertain and relies on a good mutual understanding of artistic matters. The audience, also, participates in the construction of the performance's meaning. Indeed, the novelty of avant-garde objects and the low level of their codification implies that it is hard to predict what will retain audiences' attention, unlike more codified genres, such as industry-based and traditionalist. The audience's reaction thus provides an insight into whether the experimentation was worth the work, and artists, cultural and technical intermediaries pay attention to this feedback.

Discussion

Music genres clearly drive the distribution of power over the shape of the performance's object. They define the extent to which artists, cultural intermediaries, technical intermediaries and audiences are granted the power to make this object look like what they want it to be. This power represents the level of control that they have on the script of the performance, and therefore on its social and artistic meanings. In scene-based genres, technicians have a significant opportunity to intervene in these meanings, because the power to decide is distributed almost horizontally. In industry-based genres, technicians are simply executors, and cannot directly impose their views on the artists. However, they can expect the latter to behave in a professional way, i.e. clearly communicate their meanings according to a script thoroughly programmed by them and cultural intermediaries. If technicians intervene in the script, this intervention, unlike in scene-based genres, is strictly defined in advance. In traditionalist genre, artists have more leeway to improvise and technicians must follow almost regardless of the extra work that this improvisation implies. In this genre, artists are granted significant control over the performance's meaning. Finally, technicians are creators in avant-garde genres, and must bring autonomous insights which heavily impact the script of the performance. However, artists keep the final say over the performance's meaning and are the ones who give the green light to the technicians' initiatives.

The root of technicians' power of decision lies in the mastery of their tools. If technicians resign from a concert for some reason, no music happens. Artists, audiences, and cultural intermediaries are left alone in front with tools that they cannot handle, and the use of which is necessary to make the show happen. This gives technicians a certain bargaining power, implying that they are not always obliged to act as mechanical executors. The power to act on the performance's meaning depends of course on factors related to the reputation of the technician, his/her professional trajectory, personal ties with artists or cultural intermediaries, etc. However, this power appears to be also related to the performance's genre. Beyond an aesthetic concept, Lena's music genres define the social meaning of a musical activity, which is reflected in the hierarchies, organizations, discourses of participant actors, and notably the distribution of the power to decide on the performance's meaning. Technicians transform the bargaining leeway given by their personal technical ability differently with respect to the music genre in which they work, as respecting the genre's conventions is necessary to make concerts function as re-fused performances.

Music genres act as a bridge between social and symbolic boundaries (Lamont & Molnár, 2002): they explicitly join the aesthetic shape of a symbolic good and the social conditions of their production. If music appears as it is during a concert, it is not merely related to the work of aesthetic development realized by artists or the history of music styles. The social conditions in which they are produced, in terms of hierarchies, organization, distribution of authority over artistic matters, which are described by the notion of music genres, also have a direct impact on the final result of the performance. We have seen in this chapter that some of these social conditions are related to conventions that need to be respected in order for the performance to keep its intended meaning. For instance, as traditional artists are considered as legitimate stewards of a heritage, technicians have to do everything they can to grant their demands. Conversely, in scene-based genres, technicians rarely grant eccentric demands from artists and keep a significant level of control over the performance's shape, preserving the collective dimension of the music produced in this genre. However, these conventions are not something every actor abides by as soon as they accept work in a given genre. Every actor has a specific power to mobilize in order to gain greater control over the writing of the script. The only limit to this power is the specific power of other actors: technicians' monopoly over the use of sound reproducing machines is practically limited by the fact they cannot replace the orchestra or organize the concert. Actors compose performances by incrementally negotiating their influence on the performance meaning while remaining true to the conventions inherited by previous occurrences of shows of the same genre.

These findings empirically lead one step further away from a concept of art worlds centred on artists, especially from a concept considering them as the central decision makers. Indeed, the script of a musical performance, carrying its meaning, is jointly composed by artists, cultural and technical intermediaries, and audiences. However, if the artists' role in art worlds is deconstructed, it is necessary to rebuild an explicit definition. Once artists are no longer identified as final decision-makers, the centre of the art world, and giving meaning to the whole activity, what is left for them? This question calls for further empirical investigation under a revised theoretical framework. However, from the point of view developed in this thesis, I would argue that artists are explorers. They produce sets of harmonized symbols meaningful to them and intend to transmit these sets to others, in the expectation that the latter will perceive the same meaning as them in these sets, or at least that it will trigger a form of emotional reaction. In the conception of a cultural performance, artists do the spadework in the background of cultural representations and propose a first version of the script, to which contributions of other actors will be added. The ways in which these additions are done depend on the social meaning given to the performance, and the balances of power between the different actors involved.

Conclusion

Focusing on sound engineers during the *mise-en-scène* phase of several musical performances was shown to be an efficient way to access their scripts, and to understand who has the power to make the performance's object reflect his/her vision of the script according to different artistic contexts. Scripts are made visible by the various interactions to which technical intermediaries are submitted as actors materializing the performance's object, and that are accessible to an observer. These scripts have shown that they are coherently organized by Lena's (2012) notion of genre, as the expectations of each actor shown by the observation method were coherent with the dimensions of each of the four genres. Likewise, Lena's genres efficiently explain the distribution of power over the performance's meaning among its different participants. These results confirm the validity of her approach and encourages the use of this framework in sociological approaches to music, whether by applying it to qualitative studies in order to obtain of more fine-grained understanding of script developments, or by using it in quantitative analysis of large datasets.

Lena's framework has the advantage of producing analytical categories separate from the internal struggles of musical fields, that are used for the sole purpose of social sciences. Music styles are the intake of struggles in music fields. Indeed, people can have different views on what constitutes a legitimate identification with a style. These identifications function as labelling and have consequences for the social boundaries within fields. For instance, whether a band is qualified as "pop" or "rock" or "jazz" has consequences for its membership of "highbrow", "legitimate" culture and therefore affects the standard of living of the band's musicians. By studying such a band as one of these three styles, scholars get involved in these struggles for legitimacy. They have to define their understanding of where the style's boundaries are situated in order to be able to study it. Yet, in doing so, they take the risk of reifying the style's definition and to help the people, in the field, who share their definition, by loading it with a share of cultural legitimacy attendant to academic production. In turn, Lena proposes a concept which is not involved in music fields, and that is sufficiently remote from endogenous conceptions to carry a lower risk of influence or being adopted by the fields' internal struggles.

I did not address, in this chapter, the national differences that could exist between France and the Netherlands, as I did not find patterns that would justify an analysis. Lena's model appears to be sufficient to explain differences in practice between genres, without requiring us to account for nationally induced nuances. While this framework has been described by its author as applicable to the United

States, it appears that it also applies correctly to France and the Netherlands. This transnational applicability might be explained by the fact that the music industry has followed a similar trend on both sides of the Atlantic ocean (Tournès, 2008). Therefore, Lena's framework might not be so easily applicable in contexts in which music has not followed a similar industrialization trajectory. In other words, in countries where intermediaries have been made less essential to the achievement of musical performances, and thus where the script is more directly influenced by artists themselves.

This chapter gave insights into the general conventions, according to music genres, governing the different expectations, behaviour, and influence of each actor in a musical performance. However, music genres are not the only thing defining all these rules: performances' shape and success also rest on elements contingent to the performance itself, which do not depend on genres. In the next chapter I will focus on some of these contingent elements, which concern the relational configuration of the working team and the specificities of the performance's script beyond what is carried by the performance's genre, and show how they influence potential re-fusion.

Group Engagement of Technicians and Audiences in Concerts

5

From one day to another, public presentations of the same show, performed with the same people, can have very different results. For instance, if a band plays two days in a row in the same town during a tour, one night can be dull and the other marked by a very intense collective emotional entrainment. Therefore, re-fusion is a phenomenon driven at least partially by contingent factors. Technical intermediaries are responsible for handling some of these factors: we saw in Chapter 1 that they have to prevent the environment from disturbing the implementation of the script, and we saw in Chapter 4 that they have to ensure that the performance's object respects the genre's conventions. However, if a strict respect of the latter was a sufficient condition to ensure a performance's re-fusion, the outcome of a given performance would not be subject to its actual range of variations and uncertainties. Re-fusion, in sum, is not a phenomenon that can be fully driven in an anticipated manner.

In this chapter, I will clarify how factors related to the contingent relational configuration of the technical team of a performance can influence its ability to re-fuse. In music concerts, the central position of sound technicians when it comes to materializing the performance's object, implies that the manner in which they engage in the performance is likely to influence its outcome.

Electronic supplementary material The online version of this chapter (https://doi.org/10.1007/978-3-658-33029-3_5) contains supplementary material, which is available to authorized users.

A. Battentier, *A Sociology of Sound Technicians*, Musik und Gesellschaft, https://doi.org/10.1007/978-3-658-33029-3_5

Theory

Studying music from a performance perspective is a way to study it through a paradigm that does not consider music as something transmitted from a producer to a consumer (McCormick, 2006). Instead, in a cultural performance, every actor contributes to a phenomenon of re-fusion, collective emotional entrainment, or potentially collective effervescence. Every actor is thus a form of producer of the performance's outcomes, which are: emotional insights, construction of meaningful symbols, and emergence, reinforcement, or weakening of social ties. However, we saw that participants contribute differently to these outcomes: artists bring raw content, cultural intermediaries frame them in a way that makes symbolic and economic sense, technical intermediaries shape the object targeted by mutual focus of attention, audiences provide feedback and fuel emotional escalation.

We have seen in Chapter 4 that the performance's script is negotiated among these actors, in ways that depend on the performance's genre. Nonetheless, in a performance perspective, the performance's relevance will be evaluated according to whether or not it has been able to re-fuse and to provide the emotional insights expected by participants, who get the final word about the performance's meaning. Following Collins (2004), such emotional entrainment is built on the intensity of mutual focus of attention during an interaction ritual, such as musical performances. Therefore, if we want to understand how a performance attains re-fusion, we have to understand how the mutual focus of attention is constructed before and during the performance.

Live musical performances are conceived in advance with the intention of producing various emotional outcomes. Therefore, it is possible to study how participants organize themselves in order to maximize mutual focus of attention and thus, the potential for the performance's re-fusion. In order to do so, it is necessary to distinguish, during the production of a performance, two moments that can be understood as a chain of two interaction rituals. The moment of the performance itself, of course, is easily read as such: a space is defined and organized, people identify themselves as participants, people on stage are musicians and those dressed in black, more or less successfully hiding in the backstage area are technical staff including the guy behind the mixing desk. When the music starts, everyone engages in a mutual focus of attention centred on the sound, identified as music, that comes out of the speakers, and emotional escalation potentially occurs.

The moment of *mise-en-scène* is also an interaction ritual of its own. Here again, insiders and outsiders are clearly defined: doors are closed before the official start time of the concert, and audience is not admitted before that time. The script has to be installed in the space in which the performance will take place, and in order to do so, the artists present, technical and cultural intermediaries will engage in a mutual focus of attention intending to fabricate the performance's object. This process is also subjected to emotional insights that will bring outcomes of their own for participants, who will therefore more or less identify with the symbols of the performance, feel sympathy for the other participants, get emotionally involved through the musical experience and build a desire to reiterate an artistic collaboration with participants. It is therefore a doubly crucial moment for sound technicians, in which they realize the most important part of their work, and in which they experience the interactions that will have consequences for their desire and opportunity to work with the same people or within the same artistic context. In this chapter, I am interested in understanding how mutual focus of attention is built during the *mise-en-scène*, and whether the presence of such focus during this moment influences the potential for re-fusion in the performance itself.

Considering the performance's participants as members of an organization intending to produce the outcomes of a successful ritual is a good way to do so. Indeed, it allows the use of analytical tools from the work of Metiu and Rothbard (2013), who used interaction ritual theory to solve issues in the sociology of organizations. They showed the role of what they defined as *group engagement* in producing efficient problem-solving in working teams of software developers. Group engagement is "the process by which interdependent individuals engage with each other around work tasks to develop mutual focus of attention" (Metiu & Rothbard, 2013, p. 458). In the working teams they studied, they found that group engagement is obtained through the presence of three enabling factors: high individual engagement, high frequency and informality of interactions, and the presence of a compelling project direction. Those factors increase the possibility for "task bubbles" to form during work, which are tight groups of collaborating workers highly focused on solving a particular problem, therefore living a successful interaction ritual. The intense level of mutual focus is fuelled by a shared feeling of excitement towards the task catalysed by the presence of enabling factors. This emotion, finally, is itself an outcome of interactions in task bubbles and fuels general enthusiasm towards the project.

In this chapter, I will reverse Metiu and Rothbard's perspective, and use their contribution in the sociology of organizations for a better understanding of the role of technicians in the interaction rituals that are musical performances. The teams of developers studied by them show some similarities with the teams on which I focus. Both are composed of professionals with a defined bundle of tasks whose collaboration is oriented towards the achievement of a defined project. Most problems during the *mise-en-scène* phase are technical, as its aim is to materialize the performance's object in a given environment. Therefore, group engagement within the pool of technicians is likely to facilitate problem-solving during this phase. However, while Metiu and Rothbard's teams work on a product designed to be sold to potential consumers, the "product" that is a concert is of a slightly different nature, if we follow a performance perspective. Indeed, such framing of musical events proposes that no "consumer" is present in them, as the audience contributes to the production of their meaning and emotional outcomes. Hence, the problems that technicians are confronted by are solved in order not to shape a product, but to engage another group of people, the audience, in the performance. In sum, the group engagement of people working on technical tasks during the *mise-en-scène* is expected to lead to an increase in the group engagement of people participating in the performance itself. Therefore, I will study how technicians' group engagement, reflected through their individual involvement, the frequency and informality of their interactions, and the level of compelling of the project direction, is increased during the *mise-en-scène* and the performance, and how their engagement potentially affects the audience's feedback observable from the point of view of technicians.

Method

In order to answer these questions, I will first focus on how the three enabling factors of group engagement influence technicians working on live concerts during the *mise-en-scène* and the performance. Then, I will discuss whether the implementation of such methods had an impact on the audience's feedback I observed.

Compelling Project Direction

What Metiu and Rothbard called "project direction" is inscribed in the performance's script. We have seen in Chapter 4 that audiences give different forms of feedback according to the performance's genre (Lena, 2012): vehement if they do not like the show in scene-based genres, impressive levels of excitement if they like it in industry-based genres, worshipful listening if they like it in traditionalist genres, etc. A general project direction is thus given by the performance's genre. However, we have also seen that all the actors have the possibility to influence the performance's script, and therefore the project direction. Some of these modifications are intended to increase an audience's group engagement by adding a layer of participation to the performance. I will account for this type of script effect, which can come from artists, cultural intermediaries, or the larger social context.

Frequency and Informality of Interactions

The data gathered during observations allow a detailed account of the potential for the creation of task bubbles during the *mise-en-scène* concerning technical issues. Metiu and Rothbard focused on teams working over the relatively long period of time needed to fully complete software development. Conversely, the teams on which I focus are gathered for a few hours, or a couple of days at best. They are all in the same place and therefore have numerous interactions. However, what is likely to increase the potential for colloquial problem-solving is the extent to which members of the technical team knew each other beforehand. I will test whether these ties are enabling them to better anticipate and efficiently coordinate over encountered problems.

To account for pre-existing acquaintance in technical teams, I listed their members in each observation. Then, I identified the people who were acquainted before the observation, and accounted for their ties. I identified three types of links. Participants can just personally know each other, without having really worked together before the observation. Participants can work together on a regular basis by repeatedly collaborating. Finally, a junior can be mentored by a senior worker. The number of relations is computed by counting a relation "working together on a regular basis" as 1.0, and "personally knows" and "is the mentor of" as 0.5. Finally, I computed a density of pre-existing relationships between technicians by dividing the number of ties between technicians by the number of technicians (Table 5.1). For the sake of readability, I defined three categories of analysis. Low density (coded 1) is used for cases with a density less than 0.5 pre-existing ties by technician. Middle density (coded 2) is used for cases with a density between 0.5 and 1. High density (coded 3) is used for cases with a density greater than 1.

Table 5.1   Density of pre-existing relationships in technical team for F1

	F01
Total number TI	7
Number of pre-existing ties within technical team (mentor $= 0.5$; personally knows $= 0.5$)	5.5
Density of pre-existing relationships	0.79
Low/Middle/High density?	2

Individual Involvement

I evaluated whether technicians were personally involved in the project by accounting for pre-existing ties between them and artists. I applied a similar method than the one used for pre-existing ties between technicians. In order to account for technicians' links with an artistic project and not with musicians individually, I clustered each orchestra in one unit of analysis. I counted 1 tie if the technician was working with the orchestra on a regular basis, 0.5 ties if the technician personally knows one or more musicians in the orchestra without being engaged in a regular working relationship. Then, I divided the sum of ties by the number of artistic clusters to obtain the density of pre-existing ties, and I ranked high, middle and low density using the same values than for ties between technicians (Table 5.2).

Table 5.2   Density of technicians-artists pre-existing relationships for F01

	F01
Total number TI	7
Number of artist clusters	3
Number of pre-existing ties technicians-artist clusters	3.5
Density of pre-existing relationships	3.5
Low/Middle/High density?	3

Audience's Feedback

Measuring audience's engagement is impossible to realize without a dedicated research design, which has not been done here. However, we saw in Chapter 4 that technicians pay attention to the audience and use its feedback in order to correct their actions. This feedback was discernible to me during observation of a performance, and I carefully noted expressions of enthusiasm, anger or disappointment perceptible from the point of view of the technical team, as well as direct interactions with the person behind the mixing desk. I will analyse if and how the three enabling factors impacted the presence or absence of these reactions.

Compelling Project Directions: Actors Adding Layers of Meaning to the Performance

As well as the project direction inscribed in the script by the music genre, additional features can be added to build a compelling project direction to the performance. These features intend to increase group engagement of the audience and can be placed in the script by artists, cultural intermediaries, or the general social context. These features add a layer of meaning to the performance, which pushes participants in general and audiences in particular towards a more significant group engagement.

Artists Bringing the Audience in

First and foremost, the most common way to involve the audience in the observed performances is by inciting them to dance. Dancing is a way to participate in the performance, by putting one's body in motion in rhythm with music. Since Durkheim (1912), it has been identified as a central element of rituals, as a way to attune and synchronize the body during collective gatherings. By dancing, people accompany the musicians and do not just passively watch a show. This was particularly the case in F03, F16, F17, N08 and N09, where most of the audience was dancing. In these cases, the music is explicitly conceived for this purpose, as made visible by their music styles: cumbia and semba (F16), techno (F17), electro dance music (N08 and N09), and fusion electro (F03). The audience's participation is inscribed in the artistic contribution to the script. As a result, many demands given to and by technicians focus on key musical elements which are known to catalyse the incitation to dance. In electro dance music (N08 and N09), musicians always insisted on the necessity to emphasize the bass frequencies. In F16, musicians from a cumbia band and a semba band, stressed the balance between their different instrument sections, telling the sound engineers which instruments were carrying the rhythm or the melody. They were particularly focused on who they had to hear in their monitors *in order to properly synchronize. A production assistant was even specifically required to ensure this information was communicated correctly:

> *A production assistant came with the band. She sees me at the monitor* console, and thinks I am the monitor* guy. I can't explain myself as she floods me in the explanation that the instrument of the lead singer, a dikenza, is hard to handle but crucial: "[The lead] does not want bass on it, and it does all the rhythm...well he'll tell you". I tell her that I am just an observing intern, but that I'll pass the word to the actual monitor* engineer and I point him as he is setting up the microphones on stage. She goes to him to discuss the issue. Later, the lead singer will come at the console and repeat that he does not want bass on his dikenza. (F16, Observation notes)*

These demands are common in all observed soundchecks, but they were, in the case where the audience was supposed to dance, explicitly related to the preservation of the music's ability to set people in motion. However, it is important to note that if music styles are influential here, they are still embedded in the more general music genre of the performance. Indeed, other observations had music styles that were strongly associated with dancing: F15's bands were from La Réunion and Cuba, and performed a fusion of jazz and their local styles, in which dance plays an important role. However, this music was performed in an "avant-garde" context which encouraged attentive listening, with people sitting and silently observing the band.

Audiences can also get involved through an artistic concept where its reaction drives the direction of the show. For instance, in F10, the audience was invited to loudly clap in order to influence the scenario of the play. The play was stopped regularly, and the public had to decide whether they wanted to push the play towards a happy or dramatic ending. If they clapped loud enough (which was decided by an actor on stage manipulating a hand-made clap-meter), the play proceeded to a happy ending, and vice versa. This strategy of soliciting the audience is double-edged for artists. On one hand, the audience takes part in the play and thus feels more involved than if only spectating upon it. The virtual separation of the fourth wall is diminished, which is a way to raise engagement in the public. On the other hand, if the audience does not feel involved or at least reach a high enough level of emotional involvement to participate, the result can be particularly awkward and uncomfortable. Indeed, the play can only work properly with the audience's feedback, and the situation will become awkward if they are not willing to give it. When such an artistic choice is made, technicians, like artists, have to follow the proper script according to the direction decided by the audience. For instance, in F10, someone at the mixing desk was charged with launching pre-recorded sounds which combined with the music performed by the orchestra, which changed according to the audience's reaction.

Cultural Intermediaries Bringing the Audience in

The inclusion of audiences' participation in the script by cultural intermediaries is the most explicit in N05. This observation was the concert of a quite young soul singer, whose reputation as a "must-see" in Amsterdam appeared to be completely created by a well-implemented marketing strategy. When I arrived at the place of observation, a well-known venue in Amsterdam that organized an evening for young talents, I was told by the programmer that the headline act is a rising star prominent in Amsterdam's scene. After the concert, in the sold-out venue, I asked a spectator and his friend why they came to this particular show. Both looked very surprised, and told me they did not understand how a person studying music could not have heard of this new diva who was the talk of the town for months.

However, I was myself quite surprised by the disconnect between the strong reputation of the band, and the fact that it was, as was mentioned on stage, the very first concert by the lead singer. Furthermore, the lack of experience of the musicians was clear on stage: they were quite intimidated by the experienced sound engineer who was doing their soundcheck, and they were accompanied by no less than three producers, who were mentoring them like guardians: handling the implementation of the computer-assisted music software, calling them when it was their turn to soundcheck, mediating with the sound engineer etc. Although visibly musically trained, it was very strange that these young players with such a strong reputation performed in this reputable venue, even on a young talent evening, without any previous stage experience.

The bottom line was explained to me by one of the producers, who told me that they were working with the singer for a couple of years, and that they recently implemented a strategy to get her quickly known. Once she accepted working with a manager, they released a clip on YouTube, and arranged through their professional contacts to have a very positive review of it titled simply "You need to listen to [artist name] right now" in a famous and much-followed music magazine. The clip was issued on the same day as the article, a couple of months before her planned first concert. And this is how, without a single show, the artist became the must-see of Amsterdam, and performed her first stage appearance with the status of local popstar.

The audience was genuinely enthusiastic when the artist came onstage and during her performance: the aura created through the process had been effective in driving its reactions. In this case, the audience was caught in the hype of a well-orchestrated media-driven buzz, and participated through the performance to give reinforce this reputation. This rather successful concert in a reputable venue confirms the status of the artist, which at that moment was based on nothing but the buzz. It paves the way for future important gigs, as it is now possible for the production to claim concrete realization, potentially informed by press releases of the concert and audiences' feedback on social networks. In sum, the audience here serves the production of the artist's reputation, handled by cultural intermediaries. Notably, no technician directly involved in the project was present, all technical issues were handled by the venue's team with the indifference and low aesthetic involvement characterized by collaboration with an unknown project. The key to audience engagement was however not the music's sound, but in the surrounding marketing handled by cultural intermediaries.

Society Bringing the Audience in

The French field work was done during a period in which France was under a serious threat of terror attacks. One of these attacks, on October 13th, 2015, targeted a concert venue in Paris, in which 90 people died. On July 26th, 2016, a hostage-taking occurred in a church, that ended in the murder of the priest. One observation was a concert of classical and popular music in a church. Hence, when the show was interrupted due to the presence of a suspicious car in front of the church's exit, it created a strange atmosphere with different levels of anxiety.

One of the organizers informed the audience between two sets that they should not go out through the main exit, without explaining why. Half an hour later, the chief priest of the church interrupted the show and said that the police had asked them to calmly evacuate the building as a suspicious vehicle was parked next to the exit and needed to be verified. The reactions were mixed, some people hurried to exit the building, while some members of the technical team wanted to remove the microphones on stage to avoid potential stealing. We got out last with the technical team, through the artists' entrance. People were having almost normal conversations there, but there was an unspoken desire to avoid mentioning the potential seriousness of the threat. I went around the church where I met a member of the technical team, and we reached a spot from which we could see the suspicious car, a black German saloon with tinted rear windows, parked on the pavement, right next to the main exit of the concert. It was parked about 200 meters from us, and a bomb-disposal expert was opening the car. We looked at him for a few minutes and decided to go back to the main entrance: if anything did happen, we did not want to witness it directly. We joined the audience waiting next to the entrance, and I was introduced to the priest of a church, who did not appear excessively impressed by what was happening. Some explosions happened, but they were the result of the bomb expert breaching the car's door. However, the explosions provoked a panic attack in a young person nearby, who fell to the ground and started to yell. He was calmed by the people surrounding him, and one of them told me he was a survivor of the attacks of October 13th.

After a few minutes, the police told the priest that everything was clear. The audience went back in the church, and the concert started back where it had been stopped. Visibly a bit astonished by what had just happened, the singer introduced the set by saying "Well...I never liked interludes." He was warmly applauded, and the concert started back again.

The aftermath of the attacks of October 13th has seen the emergence of a "culture of war" in France (Audoin-Rouzeau, 2017), that tended to identify the targets of these attacks as fighters defending a way of life. Although declining, this "culture of war" was still present at the time of the fieldwork. This false alarm reminded everyone present of their status as potential targets, and revived the traumatic memories of the attacks, whether they had been experienced personally or through the media. People on the spot had a shared fear which was, however, carefully hidden. The applause following the bittersweet remark of the singer can be interpreted as a way to consolidate trust between the performance's participants, as well as a way to express relief, compassion, and a will to continue the performance, nonetheless. Whatever interpretation is the right one, the event clearly accidentally attached additional meanings to the ones already designed by actors behind the fourth wall. It exacerbated some elements of the background culture which had initially not been selected to be part of the script, and which had been brought in by the particularities of the situation.

The larger social context brings its additional layer of meaning independent of the participants' role. The significance of the layer results from a common identification with an element external from the performance. Therefore, technicians, cultural intermediaries, artists and audience were impacted similarly by this unintended layer of meaning.

Informality of Interactions in Technical Teams

The density of pre-existing ties in technical teams appears to not be determined by music genres. We can also note that all cases with a high density of ties in technical teams are French, following the tendency of the French institutional context outlined in Chapter 3 to form larger technical teams. French technicians tend therefore to have more opportunities to meet each other (Table 5.3).

Table 5.3 Density range of pre-existing ties in technical teams, observations sorted by genre

	Interaction group technicians
Ref Obs	Ties between technicians
	1: <0,5; 2: between… 3 > 1
Scene-based	1: <0.5; 2: between 0.5 and 1; 3: > 1
N01	1
N08	1
N09	2
F02	2
F06	2
F07	2
F11	3
F12	3
F14	1
F16	3
F17	2
F18	2
Industry-based	
N02	2
N03	1
N05	2
F03	2
F04	2
F05	2
F19	3

(continued)

Table 5.3 (continued)

	Interaction group technicians
Traditionalist	
N07	2
F08	3
F09	3
F13a	1
Avant-garde	
N10	1
F10	2
F15	3

Following the findings of Metiu and Rothbard, the so-called task bubbles in which technicians proceed to collective problem-solving tend to emerge more frequently when the density of pre-existing ties is higher. When the density is low, technicians tend to deal with their problems alone and try to fix them without getting any help, which can take time or prevent the realization of a task:

Tonight's show is supposed to be filmed with the help of a Canon 7D camera. As NI is alone, he has to install everything and launch the record. But the stand does not have the proper screw to set the camera. NI is embarrassed: "Fuck, that was supposed to be an extra, but now I have to do it, because I told I'll do it". He tries to mount a DIY support with a music stand that he lies flat, and sets a block with two trays from the bar that he attaches with duct tape. Then he places the camera on the rather baroque set-up. However, in the middle of the show, I notice that the camera is going to switch off as the battery is low. I tell him, but he shrugs his shoulders: he is handling the sound of the show and has too much things to do right now to take care of it. (N10, Observation notes)

In this case, the technician had to handle everything by himself, and could not ask someone to find a solution to this rather simple problem. However, he was not alone, as he was working next to the light technician. He asked the latter whether another stand was available in the theatre. The light technician just answered "no" and let him sort things out with his camera. A collaboration would have been more likely to occur if the two of them were used to working together: the light technician would potentially have found a moment to help with the camera's situation, or at least would have been more involved in the solving of this particular problem. When the density of ties rises slightly and becomes moderate, problem-solving can imply several people. The longer the problem remains, the more likely it is to involve the whole technical team:

After the installation of the sound system, the stage manager* gets on stage and plugs a microphone on the routing stage box[1] to test the line. However, nothing comes out. They start to discuss about potential sources of the problem. They try different lines, but no one seems to work. Seeing that they are struggling on a problem, two interns come to see what is happening. The stage manager* explains what is wrong. The interns make suggestions, but the two others already tested their propositions. The routing box is not a simple one: it is connected to a system of routing through local area network processes. They try to work on the configuration of the box but do not manage to solve the problem. The technical chief passes by and suggest that the renting company gave them a rubbish cable, but this company is known and trusted by everyone present. However, the technical chief decides to call them. The stage manager checks the power supply to remove any doubt, and calls back the technical chief: "Don't call them! The inverter wasn't plugged, it works now." (F18, Observation notes)

In this example, more and more people get incrementally involved in problem solving as potential solutions did not provide the expected outcome, which is to obtain sound going from the stage to the mixing desk. People stopped their current activity to identify what was wrong, they made suggestions which were discussed and tested collectively up to the moment where the problem was solved. The characteristics of genre however influence the way these collaborations happen. F18 is scene-based, and therefore division of work is less marked than in N02, which is an industry-based observation:

The monitor* sound engineer goes on stage and replugs an instrument. A big "pop"[2] rings in the speakers. The audience laughs and yells a "whouuu". The front of house* sound engineer, which is also the chief sound technician for the stage, is really annoyed. Then a second "pop", louder, rings. The audience gets also louder whooing. The chief technician gets angry and sermons the monitor* sound engineer through the intercom.

---

[1] Generally, a big cable, called a multipair, links the mixing desk to the stage. This big cable is composed of small cables, each one carrying the signal from one sound source, for instance, a microphone. At one end of the big cable, all the small cables are split out in order to be plugged on the mixing desk. A "routing stage box" is present at the other end, into which each sound source present on the stage can be plugged. See a picture of the object in Appendix G

[2] When a source is plugged in or unplugged, an electrical peak is produced, as whenever any electrical contact is made or interrupted. If the line is not muted, the peak is transmitted to the amplifiers and emits from the speaker, making a typical popping sound. The sound is annoying, and can potentially damage the system, as its amplitude is beyond the normal functional range. Muting the lines when plugging/unplugging things is part of the ABC of sound engineering, which explain why their presence heavily irritated the chief sound engineer.

While the latter tries to explain himself, the chief cuts him: "I can't I'm working now"
(N02, Observations notes)

In industry-based observations, technicians are expected to fulfil the tasks they are responsible for: therefore, even though earlier in the day the two men, who were already acquainted, solved other issues together, it was not the case for this one. Muting the lines when plugging something in is a very basic move that one quickly learns to do. Having the "pop" sound in the middle of the show is too much for the chief technician, who criticises the monitor* technician without listening to his explanations.

As the density of ties within the technical team increases, people tend to solve problems in dyads or triads. Furthermore, In the observations with the highest density, teams tended to form subgroups. For instance, in F09, my respondent, the main sound technician, collected together a good share of the technical team, who were people he knew and who had also occasionally worked together. Another part of the team was employed by the venue. People there tended to form subgroups according to their role and who they are used to work with: the venue's team worked together, people hired through the main sound technician were working together, and the small team coming with rented gear also worked as a separate group. However, the principle of "my friends' friends are my friends" applied: when a technician trusts a colleague, and that colleague trusts a third person, then the third person will be trusted more easily by the first technician. For instance, it was the first time that the main sound engineer and the system engineer in charge met each other, and they had to set up and calibrate* the speakers. However, the latter was certified by the technical director* of the venue, who was also trusted by the former. Their communication was fluid, peppered with jokes, and the system engineer anticipated the needs of the main sound engineer extremely well who often thanked him for his initiatives. Their working relation was going so smoothly that I realized only several days later, when I interviewed the system engineer, that they never had met before, despite the fact that I stayed from 9AM to 1AM the next day observing the technical team.

In sum, the density of ties within the technical team increases the likelihood of formation of task bubbles accelerating problem solving. However, the formation of these task bubbles depends also on the genre's *ethos*. When the density is moderate, the whole team can get involved in problem solving. When the density is high, people tend to only need contact with one or two colleagues to resolve situations, and more easily trust people that they have never met before.

Individual Involvement

Here, also, ties tend to not be constrained by music genres, and only French cases have a high density of artist–technician ties. We can however note a slight tendency in avant-garde genre to have a higher density of this type of tie, which makes sense with respect to the important artistic insights done by technicians in this type of genre (see Chapter 4).

Existing working ties are a clear catalyser of individual engagement of technicians in the performance. In F09, for instance, the sound engineer had a long-lasting collaboration with the lead musician of the band. This tie led him to accept difficult working conditions, where he took on many responsibilities, did a working day of more than 16 hours, and accepted ambiguous contractual conditions. Despite all that, he was extremely thorough when it came to setting-up the musicians' monitors* (Table 5.4):

Table 5.4 Density ranges of pre-existing relationships between artists and technicians, observations sorted by genre

Ref Obs	Ties between artists and technicians
	1: <0,5; 2: between… 3 > 1
Scene-based	1: <0.5; 2: between 0.5 and 1; 3: > 1
N01	2
N08	1
N09	1
F02	1
F06	1
F07	2
F11	2
F12	3
F14	1
F16	1
F17	3
F18	1
Industry-based	
N02	1
N03	2
N05	1

(continued)

Table 5.4 (continued)

Ref Obs	Ties between artists and technicians
F03	2
F04	2
F05	2
F19	3
Traditionalist	
N07	1
F08	1
F09	3
F13a	1
Avant-garde	
N10	2
F10	3
F15	2

The monitor* engineer is testing the stage system, he goes around with a microphone to test the presence of feedback. The main sound guy tells him that he will help for the lead musician EQ, who does the piano and sings, because he knows what he wants. About 2 hours later, the musicians come, and validate their stage positions. After they left, the first thing done is the monitors*' EQ of the lead musician. The main sound engineer sits on his chair, speaks in the microphone and guides the monitor* sound engineer towards the proper sound: "bit less of 1kHz, bit brighter, bit less of 4k". (F09, Observation notes)

Technicians tend to accept accumulation of tasks and responsibilities when they have a strong working relationship with the artists. For instance, in F12, the sound engineer who had worked on the creation of the play's soundtrack helped to install the system, equipped actors with their wireless microphones, and played the show's sequence live. This overview of large section of the performance is an appreciated feature of the job:

"Z: So, after the creation, you go on tour to play the show's sequences? Because some people are more like: I compose the soundtrack and then, the tour, I just give it to someone else.

Yeah, yeah, indeed. But for me it's practically impossible, in fact.

Z: Oh, yeah? Technically or is it a principle?

> *No, in a sensible way, because…it's like asking a musician to play exactly what the other musician played, or to an actor to play exactly the other actor. No, that's not how things happen. The person will play it his way. […] And I have my way to do things…it's…it can really go far in the work, in the complexity…so much so that someone else cannot take my position on this in the end. We are not, we are not just fader* pushers, I don't think that's what we are, you see?" (FF, Interview)*

Being artistically involved in a project, taking an essential part in it, is a way to get one's work personally appreciated, and fuels individual involvement. On the other hand, the absence of ties with the artists can imply a form of detachment, as in F06:

> *The venue sound engineer tells me he is setting the stage for the main band, which he is also the sound engineer of. Two other bands are programmed but he tells me: "you know, I don't really care about them. I pushed for my band to be programmed here, I want their show to be good. The rest will have what is left." (F06, Observation notes)*

Apart from this lack of involvement, it can be difficult to mention something is wrong to artists if they are unknown to the technician, for instance when an instrument is out of tune:

> *During the soundcheck I notice that the guitar is out of tune. The singer/guitarist plays a chord, and I exchange a knowing gaze with the sound engineer. He tells me that for him it's a torture when an instrument is out of tune. I ask: "Are you going to tell him? – I don't know. That's a really rude thing to say to a musician". He won't tell him, but a few minutes later his fellow keyboard will take the burden on his shoulders (F04, Observation notes)*

Here, the technician does not risk overstepping the boundary of his role to correct an issue that is actually quite problematic for the sound that the audience will hear, precisely because he does not know how the musician will react to his remark.

Pre-existing ties between artists and technicians facilitate the latter's individual engagement. They know the work of musicians better and are thus able to take useful initiatives. Besides, they are more likely to accept some extra work if they are invested in the artistic value of the project. They can even claim a form of authorship if they are responsible for a part of the script writing: the perspective of having an overview and a significant part in the creative process is an important source of personal involvement. On the other hand, the absence of ties tends to limit their implication, as they potentially lack interest in the artistic project or want to avoid making mistakes around people they do not know.

Audience Feedback

Audience feedback, whether positive or negative, appears not related to the level of informality of interactions during the *mise-en-scène* or the technicians' individual involvement. The nature of observed feedback follows the lines of music genres, as described in Chapter 4: participative in scene-based genres, attentive listening in traditionalist, intense in industry-based, extremely variable and uncertain in avant-garde. Therefore, it seems that the main drivers of audience's group engagement, which in turn drives emotional entrainment and potentially re-fusion during the performance are musicians. Audiences concentrate on the music, and the meanings carried by the performance's script, before being aware of the sound's quality, which is ultimately a concern of the technicians themselves. Audiences do not react to an accurately placed reverb, a well equilibrated EQ or a properly calibrated* sound system. They react to the effects on music of the technicians' work, but with their attention is focused on the music as a whole. For instance, in the electro concert of F03, where the audience was dancing intensely, the venue sound engineer handling the live mix, who had no pre-existing ties with the band, used a trick of his own to push the audience's excitement (Table 5.5):

Table 5.5 Audience's engagement, observations sorted by genre

	Interaction group technicians			
Ref Obs	Ties between technicians	Ties between tech. and artists		
	1: <0,5; 2: between... 3 > 1	1: <0,5; 2: between... 3 > 1		
Scene-based	1: <0.5; 2: between 0.5 and 1; 3: > 1	1: <0.5; 2: between 0.5 and 1; 3: > 1	AUDPOS	AUDNEG
N01	1	2	1	
N08	1	1		
N09	2	1		
F02	2	1	1	
F06	2	1		1
F07	2	2		
F11	3	2		

(continued)

Table 5.5 (continued)

	Interaction group technicians			
F12	3	3		
F14	1	1	1	1
F16	3	1	1	1
F17	2	3	1	
F18	2	1	1	1
Industry-based				
N02	2	1		1
N03	1	2	1	
N05	2	1	1	1
F03	2	2	1	
F04	2	2		1
F05	2	2	1	
F19	3	3		
Traditionalist				
N07	2	1	1	1
F08	3	1		1
F09	3	3		
F13a	1	1	1	
Avant-garde				
N10	1	2		
F10	2	3	1	
F15	3	2		

During the concert, I am next to the mixing desk. The front of house* sound engineer has very few things to handle, as the music is pre-mixed by the musicians on stage. However, he shows me that when he feels that a drop[3] is coming, he does what he calls a "little +5", meaning that he pushes the volume fader* 5dB up. This move increases the effect of dynamic's increase when the music comes back in. As a result of both his effect and the music composition, the audience strongly reacts to the drop: people start to dance harder, yell and jump. (F03, Observation notes)

---

[3] A typical music pattern of electro music is to progressively raise tension by progressively dropping instruments one by one, then the instrument that is left holds a note or a repetitive pattern that sharply stops ("drops"). After a short silence, all previous instruments come in with important dynamic, creating the effect of a boost.

Many "drops" were present in this concert, and the audience's reaction was much stronger when the sound engineer performed this trick. However, the little push provided by the technician is not consciously perceived by the audience, whose attention is on the stage, not the rear of the venue where the mixing desk is installed. During the fieldwork, many respondents mentioned that their work was only noticed by the audience when something was going wrong, reflecting the thankless character of their position. Indeed, in a couple of observations, musicians were clearly not prepared enough to be on stage, and their performance was full of wrong notes and mistakes. In both cases, however, the sound was blamed, despite the fact that the concert's flaws had little to do with the work of technicians:

The first show, a band doing covers of famous pop songs, was out of tune for most of the show. While the second band is getting on stage, I am catching on the flight a conversation between five audience members. Someone says: "Everyone says the band was good but the problem is the sound, it is too loud...". The persons around her agree. (F06, Observation notes)

The concert starts with an out of tune cover of Angie of the Rolling Stones. There is some feedback*. The presenter comes to the mixing desk and complains about the sound. The second front of house* technician is getting in an argument with one of the concert's organizers. (N07, Observation notes)

The sound tends to become a matter of interest of the audience only when something goes wrong. Therefore, the group engagement of technicians, which tends to more efficient at problem-solving during the *mise-en-scène*, is not directly a driver of the audience group engagement in the performance. Instead, the audience group engagement appears to be driven by the performance script implied by the genre, and the eventual layers of meaning added by artists, cultural intermediaries, and the larger social context. During the performance itself, the audience's group engagement is handled by artists, as the former's attention is addressed to the latter. Technicians do play a role in this engagement, but the efficacy of their actions remain undetected, not noticed by audiences unless something goes really wrong. In sum, the better the technicians work, the higher their group engagement is, the more they are not noticed by the audience.

Discussion

Group engagement of technicians appeared to be a significant enhancer of technical ability. Indeed, the presence of informal interactions and individual

engagement impact whether problems technicians are confronted with are solved. Depending on engagement, they become more or less able to materialize
the performance object and to perform the various tasks related to their position.
However, once the *mise-en-scène* is done, the ability to engage audiences and thus
to re-fuse the performance is mostly held by the musicians on stage, who must
perform according to the script, music genre, and the eventual specific meanings
brought in by themselves, cultural intermediaries or society at large.

The findings in this chapter show the mechanisms by which technicians relationally become more able to address complex problems during the *mise-en-scène*.
While they work in short-term projects within constantly changing teams, previous knowledge of each other is a way to obtain better results. Furthermore, a
close relation between them and artists is also a way to improve efficient problem-
solving, as they are more able to anticipate potential problems, and are more
personally invested in the performance. The difference in institutional settings between France and the Netherlands have therefore a direct effect on the efficiency
of technicians' work. We have seen in Chapter 3 that a strong social protection
system and a more subsidized context encourages the recruitment of more aides
and assistants. In these types of positions, people have the opportunity to meet
each other and to learn how to work together. As a result, they become more
efficient when they work together again, in another time and place.

However, it appears that technical ability does not directly contribute to
audiences' group engagement, which is fundamentally determined by the actions of artists, and their coherence with what is expected by audiences according
to the type of script that has been promoted by cultural intermediaries. The work
of technical intermediaries is more likely to be noticed if there is a problem:
the sound is so distorted that it cannot be identified as the music of the band,
the musicians play but no sound comes out of the speaker, the lights go out
etc. This aspect of their work has been noted for other technical intermediaries,
such as translators of TV shows (Kuipers, 2015). In music performances technical
intermediaries actions on sound can have a direct effect on the audience, beyond
the plain fact that the concert could not happen if they did not set up the gear
necessary to play it. We have seen that some of their actions, like tweaking the
volume at the right moment, affect the audience's group engagement. However,
these actions are perceived as resulting from musicians, not technicians.

This finding reflects a more general limit to the role of intermediaries in art
worlds. Indeed, artists rely on them for a very important share of their activity. Cultural intermediaries make and break careers, frame artistic content for
audiences, use their expertise to write performance scripts that make sense,
influence the direction that cultural production takes (Lizé, 2016; Maguire &

Matthews, 2012) and so on. Technical intermediaries produce an object that reflects both artists' creations and cultural intermediaries framing, and this object will be the only one that will be ultimately perceived by audiences. However, the latter holds artists responsible for re-fusing the performance. Despite the important and growing role of intermediaries of cultural production, artists are always, as authors, the ones providing the "compelling project direction" that brings audiences in. This direction is collectively conceived, but those on stage are the ones carrying it through an embodiment of everyone's work. Therefore, they are the ones getting the fame or the shame depending on the performance outcome, i.e. its ability to engage audiences in the production of emotional outcomes and their bonding effects.

I would argue, following McCormick (2006), that a performance perspective is an efficient way to understand music as a performing art. In sum, what mostly differentiates a performance perspective from a "production/consumption" perspective is the role played by audiences. Following this theoretical framing, I analysed the role of the audience as a producing actor, not consuming one. The theoretical tools applied to teammates in production teams appeared to also apply to audiences in a performance perspective, which indicates that they can be empirically approached as producers. This result must encourage studying music, as well as other forms of performing and non-performing arts, in this way. Indeed, it gives us the opportunity to understand such activities in the light of their ability to produce emotions and social ties, which is likely to constitute an exciting and promising avenue for interesting research.

Conclusion

In order to re-fuse, performances mostly rely on the coherence between what is shown to the audience and the script that they are supposed to follow. The actions of participants are made with the objective of achieving this coherence. This is the case for sound technicians, who negotiate and adapt their actions and decisions according to the needs of each type of actor regarding the performance's object. However, their ability to do so, or their technical ability, appeared to be heavily influenced by the relational configuration of the team they were working in. Indeed, pre-existing ties between technicians facilitates mutual assistance and informal problem-solving. Besides which, pre-existing ties between technicians and artists facilitates mutual communication of essential information and increases the technicians' personal involvement.

However, even though the group engagement of technicians is essential to proper management of the materialization of the performance's object, it appeared to be not directly related to the group engagement of the audience, which is handled by artists building up on the work that cultural and technical intermediaries have done beforehand. I have shown the importance of technicians in music performance throughout this dissertation and we finally reach their limits here. As intrinsic as their participation can be, and despite the fact that they are ultimately shaping what will be heard by audiences, a performance re-fusion depends on what the musicians provide, and not on the way intermediaries provide it. Therefore, the growing importance of intermediaries must not overshadow the primacy of artistic insights. Intermediaries only shape and frame creative impulses, and despite the importance of shaping and framing, emotional outcomes provided with the presence of an audience rest first and foremost on the work of artists themselves.

Conclusion: Technical Ability, Art Worlds, and Performance Perspective

When I arrived at the university after my high school degree, my goal was to become a sound engineer. However, after a couple of years of training experience and a year of work, it seemed to me that a master's degree would be a nice opportunity to try to answer some puzzling things that I had been confronted with. Step by step, I decided to pursue a PhD with the intent of focusing on technicians. While presenting my research project in order to find supervision and funding, I have been confronted more than one time with a question I did not expect: "who are these people? Who are you calling "technicians"?". As I worked and navigated for more than two years around them, I had become aware of the existence of a group, defined by its professional occupation, called "technicians". I had a fair idea of its bundle of tasks, and I experienced being identified as one of them in different artistic and social contexts. In sum, I knew that "technicians" exist as a social group. However, it appeared that it was not so self-evident through a sociological perspective.

This tension, as I understood later, was caused by the fact that existing research on the population I intended to study was rare, scattered, and not unified in a common theoretical architecture. Despite the works analysing art as a collective construction (Becker, 1982; Lizé, 2016; Negus, 2002), framing objects and technologies as active agents of social life (Hennion, 1981; Latour, 1990; Law, 2009), and focusing on the various institutional settings influencing cultural production (Bourdieu, 1971; DiMaggio & Powell, 1983; Grisworld, 1986; Peterson & Anand, 2004), sociology had not yet examined the role of the people handling technical matters in art worlds. The notion of technical intermediaries results therefore from the necessity to first and foremost name my research object and to be able to place it in the stream of theoretical constructions in cultural sociology. However, formulating this concept did not answer another provocative remark I have been confronted with: "Well, if nobody has been interested so far, maybe it's because it's not interesting?".

Technical Ability as the Governing Principle of Technical Intermediaries

Good point. However, I would argue that this dissertation proves that understanding how technicians contribute to the production of an artwork brings many interesting sociological insights. The first one is a better understanding of how various meanings come to be embodied in, and mediated by, a physical object. Technical intermediaries in cultural production ensure the embodiment of meanings in a piece of material that can be measured in physical scales of space and time, and thus can be manipulated. I focused here on how sound engineers work to ensure that variations of air pressure, which is the way to characterize sound in the terms of physics, effectively carry a system of meaningful symbols identified as music. To answer this question of how they work, I benefited in particular from the insights of Dominguez Rubio (2012, 2014, 2016) who had already extensively showed the existing tension between the physical form of an object and its ability to be perceived as a piece of art. By focusing on how an art institution, the MoMa in New York, deals with the mission to preserve the aspect of the pieces within their walls as time and oxidation undermine and degrade living and non-living things, he showed that the qualification of some objects as art depended to a certain extent on the preservation of some of their key physical characteristics, and on the presence of a will to engage the means necessary for this preservation (Domínguez Rubio, 2014). In his latest piece, he identified conservators as key actors of this preservation, building on an analysis of the work done to conserve the *Mona* Lisa (Domínguez Rubio, 2016), and I identified these actors as technical intermediaries.

However, Rubio arrived at technical intermediaries through an object-oriented approach informed by STS theoretical frameworks, assuming in particular the generalized symmetry of actions of objects and humans. Therefore, his analysis placed agency on the objects' side, and humans are always trying to keep up with the conditions imposed by these objects. The technical intermediaries who were conservators in his analysis were simply applying various techniques and technologies intending to counter objects' degradation: their leeway is defined by the amount of technical means they have access to, means that are incarnated in the tools that they use to do so. By studying technical intermediaries through a human-based approach, I showed in Chapter 1 that this access to technology and the challenges posed by objects was one parameter among others of the ability of technical intermediaries to give cultural objects their meaningful shape or to maintain it, which I called "technical ability". Moreover, I argued that this non-relational parameter, resulting from the physical characteristics of objects and environments, explains only a minor part of technical ability, compared with the relational parameters that I explored in the following chapters.

In Chapter 2, I explored a fundamental aspect of technical ability: the access to the economic means necessary to gather both the tools and qualified personnel to shape a cultural object. I focused on live pop music performances, and I showed that cultural policies and conditions of employment crucially shape access to the resources necessary to materialize such performances. I started from the demons-tration of Baumol and Bowen (1966) that performing arts' economic viability is threatened in societies in which some sectors' productivity is skyrocketing. The mechanism of Baumol's law implies a constant increase in personnel costs, and therefore a constant increase in the cost of hiring technicians to materialize per-formances. Technical ability, in this case, relies on the ability of music venues to pay the qualified personnel who will make the concert possible. I identified that in France and the Netherlands there are two ways to keep venues able to do so. Public money can pay for the personnel through subsidies, with the side effect of giving public authorities a form of control over the venues' activity; or venues can lower their personnel costs by externalizing their workforce.

The latter solution has a direct effect on technical ability in a given institutional context, as it hardens the terms of the technicians' labour market, degrades their financial situation, and potentially drives people towards professions in which a better standard of living is ensured, thus diminishing the pool of available pro-fessionals capable of efficiently delivering a concert. However, it is possible to counter this effect by managing to restore the stability of technicians' financial situation, that will maintain a pool of competent professionals in the labour mar-ket. I showed that it is what the French system of unemployment for freelancers in performing arts does. I argued that the money transfer from highly productive sectors somewhat balances out the structural deficit, that Menger (2015) percei-ved as a flaw in the system, and was in fact one of its best qualities. This transfer can be understood as a redistribution of the benefits of productivity gains aiming to restore the economic viability of performing art activities. The benefit of the French system is that the economic sectors in which production practices trigger the effects of Baumol's law pay for its consequences through the unemployment welfare of freelancers in performing arts. Therefore, the people who increase the pressure on personnel costs contribute at the same time to reduce it. In doing so, they contribute to raising the bar of technical ability as this redistribution reinforces venues' ability to hire the personnel necessary for stage performances.

However, this is not the only effect of the differences in institutional contexts on technical ability that I had the opportunity to observe and describe. One of the effects of loosening the conditions of technicians' labour market is to create a larger number of positions for assistants and aides in the personnel gathered for staging musical performances. Chapters 3 and 5 discuss how these positions directly increase technical ability at a team level as they allow technical teams to cover a larger number of issues, and to cover them more efficiently by facilitating communication within teams. Furthermore, these positions are an important vector of on-the-spot training: technicians hired to fill them learn skills and practices that they will be able to re-use in other contexts, as well as becoming familiarized with the artistic conventions of a plurality of art worlds. These transferable skills, are also an essential tool for a technicians' professional mobility, as they allow them to travel between art worlds and find more job opportunities.

When assistants and aides are present, the tasks related to the performance's materialization are distributed between: on the one hand, technicians taking decisions aesthetically influencing the shape of the performance's object, i.e. the thing that one can describe in multiple numerical scales which will be the target of the performance's mutual focus of attention; and on the other hand people who mostly handle the performance's staging without having to take these kinds of decisions. The latter play an important relational role, as their tasks are shown to be at least as relational as material. While assistants and aides take charge of issues related to material matters, such as setting up the sound system or placing barriers to define the space into which the audience is permitted, they also handle the task of facilitating communication between the different participants. As respondents put it themselves, a good aide is an aide who anticipates his/her colleagues' needs. This implies significant familiarity with the communication rules in their workplace, which change between art worlds, and that one must get accustomed to in order to efficiently guide everyone's behaviours and actions towards the realization of the performance. For this, a practical knowledge of everyone's needs and *ethos* is necessary. This is even more salient for technicians who have to take aesthetic decisions, who need to be accustomed to the specific requirements of the music they are working on, as well as to the specific needs of the musicians who perform it. Therefore, here again, I must emphasize the relational dimension of technical ability: knowing how to handle machines in order to shape sound is far from being sufficient to do a technical job well.

The fact that technicians can influence the performance's script by taking decisions regarding the aesthetic dimensions of the object they are shaping contrasts sharply with describing them as "support personnel" (Becker, 1982), which they are usually labelled to as. It goes against the claim that such personnel do not make "choices that give the work its artistic importance and integrity" (Becker, 1982, p. 77). That said, it is necessary to understand how these choices are distributed. I showed in Chapter 4 that the notion of music genre, as understood by Lena (2012), was an accurate way to understand how the power to influence the performance's script was distributed among participants. I empirically verified the presence of typical hierarchies, division of work, and decision-making processes that are coherent with the dimensions of each of the four genres she proposes. I argued that the accuracy of her model lies in the way the concept was construed. By mixing insights related to the symbolic embodiment of music styles with social properties of the actors involved in it, her notion of music genre efficiently bridges social and symbolic boundaries (Lamont & Molnár, 2002) and allowed me to look at music activities from a new perspective.

Regarding power relations over the performance's meaning, technical ability relies on the ability of the actors present to reach an agreement upon the shape of the performance's object. Music genres provide a guideline that roughly dictates hierarchies in this matter, but all actors have some leeway that they can use in order to impose their view when opinions diverge. Technicians' leeway comes from the mastery of their tools which they are the only ones able to use. In an extreme case, if they refuse to work, nothing happens: the performance cannot be materialized. It stays at the "un-incarnated" script phase and nobody will, literally, hear it. However, they are limited in how they can apply this power by the genre's conventions and the specific powers of other actors, which are related to the extent to which other roles are needed in the performance: technicians will not take to the stage and play for the band, or organize the concert. I did not explore, however, all the various ways in which power over the performance's meaning can be distributed, and it is a certain that many others exist.

If technicians have the power to block a performance, they can also get particularly involved in it. I studied in Chapter 5 how relational factors were able to facilitate their group engagement (Metiu & Rothbard, 2013), and how this group engagement was likely to reflect on audiences. Even though the presence of pre-existing ties, whether between them or with artists, was likely to improve their technical ability by increasing the capacity of a working team to efficiently collaborate on problems encountered during the performance's *mise-en-scène*, I was not able to find any link between technicians' group engagement and the level of feedback provided by audiences. I saw in this absence of link the limit of technicians' contribution to musical performances: re-fusion is ultimately achieved by the actions of artists. Intermediaries, whether cultural or technical, handle the necessary conditions in order to facilitate the occurrence of this phenomenon, but only artists bring sufficient conditions for it to happen.

If it had to be summarized in one sentence, the contribution of sound technicians, as technical intermediaries of musical art worlds, could be described as: creating the material conditions necessary to the achievement of a performance's re-fusion. They do so by ensuring the preservation of a system of meanings is retained when implemented in a material object. They potentially participate in the conception of this system of meanings, or just ensure that artists, cultural intermediaries and audiences find what they expect from this object. They help to find a consensual agreement on the performance's object by relying on relational skills gleaned from experiences in various aesthetic and social contexts in which they have navigated. This reading, however, is one of the many that are possible on technical intermediaries, and brings the material element into the theoretical framework of cultural performances and interaction ritual chains, which explain how emotional insights tied to various forms of human encounters build social bonds which combine to form the meanings, identities, and beliefs upon which societies rest (Alexander, 2004; Collins, 2004). While these approaches emphasized the role of the mutual focus of attention for building these bonds, they paid little attention to the material foundations of this focus. However, emotional escalation cannot occur without some sort of material traction. In musical performances, this traction is handled by technical intermediaries, and therefore studying them has been a way to shed light on this link hidden in the theory. In this regard, technicians are a promising research object.

Theoretical Contributions of this Thesis

Beyond explaining how technicians contribute to music worlds from a performance perspective, this dissertation proposes an operational concept that can be used to study technical work in art worlds, everywhere where "a production system meets the vagaries of the material world" (Barley, 1996). The concept of technical intermediaries completes the existing works of intermediaries of cultural production, that so far mainly focused on people handling the symbolic capital of artworks, performances and artists (Lizé, 2016; Maguire & Matthews, 2012; Negus, 2002). The absence, in the works on cultural intermediaries, of "sound engineers, camera operators, copy editors and so on" was noticed by Hesmondalgh (2006, p. 227) who proposed covering them through the concept of a "project team". However, this notion might make the mistake of framing artistic production as a collective activity, which blurs the actual roles that actors empirically take, as well as differences in authorship, and therefore give a hard time to anyone who would like to untangle these roles and how they can be present from one art world to another. Wright (2005, p. 110) proposed differentiating "makers of meaning" that would be cultural intermediaries from "makers of things" that I identify with technical intermediaries. Although this distinction functions in principle, it can be misleading in practice. Cultural intermediaries' do have an impact on things, such as contracts, programs, venues, posters etc. Certain types of cultural intermediaries, for example record producers* (Hennion, 1981) have a direct impact on artistic content, as they can closely drive artists or technicians in their work. On the other hand, technical intermediaries, as we clearly saw throughout this dissertation, also have an impact on artworks' meanings. Therefore, it is necessary to nuance this distinction between makers of meanings and makers of things, which is not as clear cut as the formulation might suggest. The difference, in fact, resides in the object of their attention: technical intermediaries *are focused* on things, while cultural intermediaries *are focused* on meanings.

The identification of technical intermediaries as a category of analysis has led me to model art worlds in a slightly different way than that proposed in Becker's essay (Figure 1). In Becker's model, the group of people that make the artwork appear *as it is* surrounds the artist. Every step is ultimately there to serve the artists' creativity. Therefore, in Becker's model, everyone who is not an artist is ultimately support personnel. The labels of other analytical categories are only there to qualify the way in which they give their support to artistic activities, in which the artist is central. The label "support personnel" just qualifies what cannot be qualified without questioning the centrality of artists in art worlds, which explains why it is described as a "miscellaneous category designed to hold whatever the other categories do not make an easy place for" (Becker, 1982, p. 2)

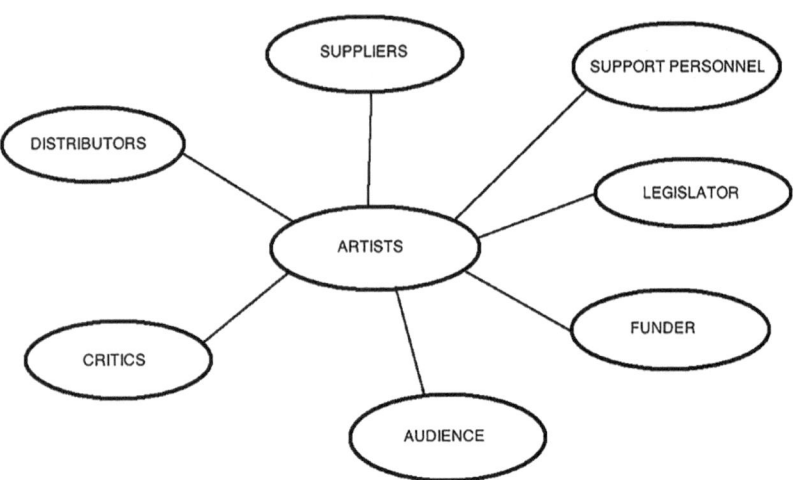

Figure 1   Arts worlds' model according to Becker (1982)

However, as Marcel Mauss put it in 1936, miscellaneous categories are often the ones in which new things can be discovered:

> There is always a moment, as the science of certain facts is not yet condensed in concepts, those facts not even being organically grouped, when we plant as a picket on these masses of facts the mark of ignorance: "Miscellaneous". This is where you have to fathom. We are sure that it's here that some truth is to be found: first, because we know that we do not know, and because we have a vivid sense of the quantity of facts. (Mauss, 1936, p. 271)

Focusing on support personnel has led to a re-modelling of art worlds in a way that removes the miscellaneous category (Figure 2).

This model reminds us of Griswold's "cultural diamond" (Griswold, 2008) but that is only because they are both the same shape. However, unlike the cultural diamond, its poles strictly contain people, and the model therefore encloses the various categories of actors that can be found in the production of a cultural object. Art worlds are represented as a clique of interrelated and interdependent actors, of which nobody is the centre. This model is particularly relevant from a performance perspective: if the point of artistic or musical performances is to collectively build emotional insights, there is absolutely no reason to put artists at the centre of the process. Indeed, if one of the actors, artists or other, does not play

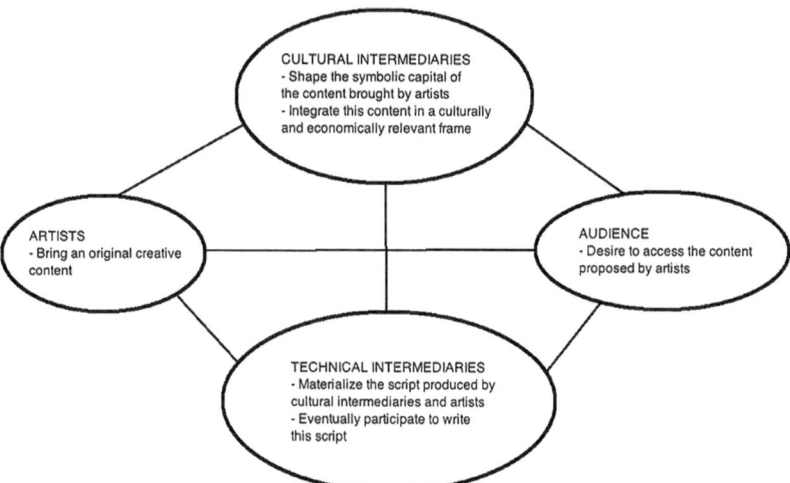

Figure 2 Art worlds' model used for this dissertation

their role, then the performance cannot happen. Therefore, each actor is as central as the other, which is a textbook definition of a clique (Zweig, 2016). Besides, in this model, intermediaries are not bringing artists to an audience, but are organizing the reciprocal movement of one towards the other. They do not provide the music to audiences: that is the role of artists. But they organize the encounter between them. If intermediaries of cultural production were a dating agency, cultural intermediaries would orient people towards their match, and technical intermediaries would decorate the room in which they will meet.

I would argue that this model brings a better understanding of what makes artworks happen as they are, because it analyses their production as the result of a process in which every actor brings a specific but equally necessary contribution. This model works for analysing performances, as has been proved for this thesis. However, it can also be used to analyse the production of artworks not based on a performance, by analysing the social and emotional outcomes of the interaction rituals or cultural performances in which these objects are involved. Indeed, the meaning of art objects such as paintings or sculptures is also built during interaction rituals in which the four types of actors listed above are involved.

Therefore, it must be emphasized that these meanings are built through the social relations between people in contact with these objects. The objects in themselves, even if they support these meanings, do not play an active role in their construction. I argued in Chapter 1 that studying the work of sound engineers, as technical intermediaries, brings important questions regarding generalized symmetry (Callon, 1986; Latour, 1990; Law, 2009). Objects can be understood to be active agents of the social construction of meaning only as long as technicians stay under the radar of sociological analysis. Their work is black-boxed and identified as resulting from objects' active role. However, once the black-box is opened, the effort realized to integrate objects and environments in an actor network appears to be driven by mostly relational factors. The object's "contribution" is limited to the establishment of a constraint, that relationally constructed technical ability will overcome or not. I conclude that objects cannot be understood as agents having as much leeway as humans in the construction of society. This construction must be explained by factors related to relations between humans, objects having a real but only minor influence in this explanation.

In this thesis, I developed a theoretical method allowing the inclusion of materiality in the analysis of art worlds without relying on the principle of generalized symmetry, by borrowing from interaction ritual analysis, cultural sociology, institutional analysis, and sociology of organizations. Object-centred approaches, such as STS and ANT, have shown social scientists that as human societies evolve in material environments, they cannot be understood without accounting for them. Humans build complex relations with their material surroundings, and dismissing them prevents us from properly understanding what society is. However, considering materials as the central objects of social analysis leads to another pitfall, ignoring conscious and regulated distribution of power among humans. This critique has been given by authors quoted in this dissertation (Elder-Vass, 2015; Lettow, 2017), and its relevance is exemplified in the current manuscript. Indeed, while sound-reproducing technologies have changed many things in the way music is produced and consumed, they have not changed the basic structure of it. Sound technicians studied in this piece are responsible for bringing music to the ears of the audiences. They probably share this concern with ancient Greeks and Romans responsible for designing theatres, who incrementally improved their acoustics through time (Chourmouziadou & Kang, 2008). Therefore, the revolutionary changes that a new technology can bring in society have to be relativized, as it just affects the surface of a social process and not its core principles. For instance, we have seen in Chapter 5 that despite the growing importance of intermediaries' work, whether technical or cultural, artists remain ultimately the people on which the success of the performance rests.

I have shown in this dissertation that sound-reproducing technologies are used in a variety of ways, which follow the boundaries drawn by notions issued from a human-centred approach of society: national boundaries, education, hierarchical organization, music genres, density of ties, etc. In other words, these technologies do not produce their own social world, as much as they influence the configuration of the existing one. Interaction rituals chains are the mechanism to allow the understanding of the mechanisms of this influence, because it provides a framework for a governing principle leading people to engage in an interaction with their fellows. From there, one can read art worlds as social constructions built to satisfy basic needs in dedicated interaction rituals. Insights from various theoretical frameworks can be integrated to understand how these constructions are built, evolve, and disappear.

To conclude, I argue that the renewed gaze brought on the mechanisms of Baumol's law (Baumol & Bowen, 1966) is a good example of the kind of analysis that such an approach fosters. The formulation of this law strongly defines performing arts as an archaic sector, incapable of keeping up with the wheel of progress and its productivity increases. The term "cost disease" frames it as sick due to its inability to increase its productivity. However, recent works have shown the importance of this "sick sector" and its creative insights in capitalist economies (Boltanski & Chiapello, 2011; Florida, 2005). However, an activity cannot be at the same time archaic and avant-garde. I propose a different framing of the economic relation demonstrated by Baumol and Bowen. The relational work necessary to properly stage a performance does take an uncompressible amount of time which no technology of mass production is able to reduce. But this cannot be interpreted as a sign of archaism, otherwise any kind of social relation such as love or friendship could be called archaic if the time necessary to achieve it cannot be reduced through some kind of technological process.

Instead, Baumol's law must be understood as the demonstration of an incompatibility between constant production growth, which the authors incidentally never question in their demonstration, and the economic viability of performing arts. The cost disease is a choice, not an inevitability. It is a choice made at a macro-level, a choice of society at large, that is expressed on a micro-level through how productivity gains are used. Are they used to raise production volume or to diminish labour volume? If they are used to diminish labour volume, how does workers' income evolve? Are they guaranteed a reasonable income? If they are used to increase production volume, how are the activities that cannot keep up and are therefore condemned to an economic death treated, burying the specific knowledge they carry and the social ties organized around them?

The performing arts, in this regard, are a lucky exception: they have been historically considered worth running at a loss. One can argue that some activities are not worth keeping when technological progress is on its way. However, progress will probably experience a change of direction in the next decades due to the radical changes that the ecological crisis will bring (Bendell, 2018). The role, in this crisis, of economic growth, notably fuelled by the systematic use of productivity gains towards a rise of production volume, has been highlighted, and its responsibility is the topic of intense and crucial discussions (Jessop, 2012). Therefore, the "disease" identified by Baumol and Bowen might arguably not be in the costs of performing arts, where it was initially thought to be, but at the other end of Baumol's law, in the constant use of productivity gains for the purpose of increasing production volumes.

Limits and Perspectives

This dissertation did the spadework for further studies of technical intermediaries, for the role and relations of various types of actors in art worlds, and for studies of performances. It developed new concepts and gave various insights into their role in art worlds. However, although it led to these theoretical developments, it relied on a limited number of respondents (17 in France, 11 in the Netherlands, and 1 in Canada) and observations (30 observations in total). Furthermore, the fieldwork was relatively short, and diversity of cases studied was preferred over depth of ethnographic scrutiny. Therefore, its insights can be refined and developed through an extension of the respondents' panel, as well as longer and closer observations. Besides, it has mostly focused on live musical performances, as it was the easiest way to perform direct observation of technicians' work. The work of sound technicians, and more generally technical intermediaries, is likely to take other forms in, for instance, recording or in other artistic contexts.

Technicians' careers are a promising object of research. The diverse nature of their professional engagements interrogates the role of transferable skills in project-based labour markets. Furthermore, the mobility that they have between art worlds is likely to grant them a role in the distribution of conventions and practices fuelling artistic innovation (Uzzi & Spiro, 2005), which I have not been able to properly untangle in this dissertation. The way they oscillate between "technical" and "artistic" positions and the extent to which they are granted authorship are also likely to be the source of useful insights into the organization of cultural production and the extent of power over the meaning of a cultural object granted to technical intermediaries.

The model proposed here to describe art worlds gives the possibility to easily operationalize a systematic study of the distribution of this power. In Chapter 4 I developed a promising step in this direction, which can be followed by many others. Even though I showed that music genres are useful for understanding who decides the shape of a performance's object, and ultimately, its script and its meaning, the mechanisms of this distribution still need to be untangled in more detail. It is notably a way to understand the nature, extent, and limits of the influence of cultural intermediaries across time and in different contexts of cultural production.

Another potential expansion of this dissertation is to further explore the relations between the group engagement of people involved in a performance, and its actual re-fusion or the level of collective emotional entrainment it triggers. Even though I have not been able to show a direct link between the group engagement of technical intermediaries and audience's engagement, I remain convinced that a high level of artists and intermediaries' group engagement exerts, potentially indirectly, an attracting effect on audiences, which results in exceptionally successful performances. Metiu and Rothbard's (2013) framework of group engagement provides an excellent operational way to explore such questions in a study that, unlike this one, would pay more analytic attention to the effects of performances on audiences.

To pursue these potential developments, qualitative comparative analysis was shown to be a very efficient tool of analysis. Indeed, it has allowed me to investigate an important amount of complex data in a way that highlighted important relationships within it, especially in Chapters 3 and 4. I did not find many works, based on ethnographic data, that used QCA as an analytical tool. Its efficiency in this dissertation encourages greater use of it. However, even though it has been a very good way to untangle important information through the jungle of interviews and fieldwork notes, I have not directly used it to support theoretical claims. I did not use logical minimization to produce a general statement answering my research questions, but always interpreted the ties that QCA revealed using a variety of theoretical tools. It seemed to me that logical minimization was not sufficient in itself to provide definite answers, and that focusing on intermediate indicators, such as consistency scores in Chapter 3, was a better way to interpret tangible information. QCA, as a method of analysis of ethnographic data, has a lot of potential and deserves to be developed, although there is still some way to go before it can attain the level of mathematical robustness of probability theory.

I conclude by emphasizing the benefits of the performance's perspective. As McCormick (2006) pointed out, choosing this approach is choosing a paradigm in which audiences are producers of a performances' outcome, just as much as every other actor involved. Such a perspective on cultural phenomena is a way to further understand them as resulting from collective actions, rather than a transfer between two parties. It does so while remaining able to efficiently account for the power dynamics and nascent hierarchies at stake during a cultural performance, as well as for environmental factors. Its ability to do so relies on the fact that it focuses on the moments when meanings and social ties are produced, in the context of the various pressures introduced by existing hierarchies. It has therefore a good balance between agency and determinism: it studies social change (Summers-Effler, 2002) in a social context bringing a range of open possibilities. For this, it deserves to be more widely used as a paradigmatic approach.

Glossary

- Agent: cultural intermediary who finds and negotiates job opportunities for an artist.
- Art director: cultural intermediary driving the artist towards a form of expression which fits his/her aspiration and distribution needs.
- Backliner: technician handling music instruments.
- Booker: cultural intermediary who finds gigs for a musician.
- Calibrate the PA system: work the EQ and others setting up the PA system in order to ensure that the sound is not distorted by it.
- Capture: capturing of the sound of an instrument or a voice.
- Devolution act: "decentralisation" in French. Policy intending to redistribute institutions and powers throughout France, as they are concentrated in Paris.
- Editor: technical intermediary selecting and assembling sound and/or image rushes to make a film/record.
- Fader: line knob, often used to balance to levels in a mix.
- Feedback: also known as Larsen Effect, strident sound happening when a source and speaker are looped.
- Festival: I noticed a difference in the term use between France and the Netherlands. In French, it is exclusively used to describe a several-day or exceptional event. In the Netherlands, it can also designate a thematic evening with several bands (for instance, country or electro bands).
- Front of house sound: sound coming from the front speakers, directed to the audience.
- Microphone sensitivity: property of a microphone, consisting of more or less important transfer of the amplitude and grain of captured sound.
- Monitor: speaker placed on the stage, directed at a musician in order for him to hear him/her-self and the rest of the band.
- Producer: vernacular term defining various forms of collaboration with musicians in the process of producing a recording.

- Rehearsal service: in classical music, a service is a time in which musicians play in orchestra.
- Resonate: application of a frequency due to a phenomenon of interference.
- Restitution: transformation of an electrical signal into sound by a speaker.
- Road crew: technical crew accompanying the band on the road. By extension, the technical team.
- Roadie: member of the technical crew taking care of manual tasks.
- Sound design: shaping the soundtrack of a performance. For instance, a theatrical play, an escape game.
- Sound engineer/technician: the two terms are practically equivalent, although "sound engineer" tends to be more appropriate for the manipulation of mixing devices, while the technician term fits better for someone less oriented towards aesthetically influential tasks.
- Stage manager: person responsible for the placement and timing of objects and events supposed to be placed and occur on stage
- Stagehand: technical intermediaries handling manual tasks and system plug.
- Technical director: the head of technical team, in charge for the gear and personnel's proper deployment.

References

Abbott, A., & Tsay, A. (2000). Sequence Analysis and Optimal Matching Methods in Sociology: Review and Prospect. *Sociological Methods & Research*, 29(1), 3–33. https://doi.org/10.1177/0049124100029001001

Alexander, J. C. (2004). Cultural Pragmatics: Social Performance between Ritual and Strategy. *Sociological Theory*, 22(4), 527–573. https://doi.org/10.1111/j.0735-2751.2004.00233.x

Akrich, M. (1992). The De-Scription of Technical Objects, In W. Bijker & J. Law, *Shaping technology/Building Society: studies in sociotechnical change*. Cambridge, The MIT press.

Akrich, M. (1993). Les formes de la médiation technique. *Réseaux* 11(60), 87–98.

Akrich, M. (2010). Comment décrire les objets techniques? *Techniques & culture*, 9.

Akrich, M., Callon, M., & Latour, B. (Dir.) (2006). *Sociologie de la traduction: Textes fondateurs*. Paris, Presses des Mines.

Ashford, S. J., Caza, B. B., & Reid, E. M. (2018). From surviving to thriving in the gig economy: A research agenda for individuals in the new world of work. *Research in Organizational Behavior*, 38, 23–41. https://doi.org/10.1016/j.riob.2018.11.001

Audoin-Rouzeau, S. (2017). L'après-13 Novembre : Naissance et mort d'une « culture de guerre » ?. *Vingtième Siècle. Revue d'histoire*, 134(2), 11. https://doi.org/10.3917/ving.134.0011

Barley, S. R. (1996). Technicians in the Workplace: Ethnographic Evidence for Bringing Work into Organizational Studies. *Administrative Science Quarterly*, 41(3), 404.

Bates, L. J., & Santerre, R. E. (2013). Does the U.S. health care sector suffer from Baumol's cost disease? Evidence from the 50 states. *Journal of Health Economics*, 32(2), 386–391. https://doi.org/10.1016/j.jhealeco.2012.12.003

Baumol, W. J., & Bowen, W. G. (1966). *Performing arts – the economic dilemma: A study of problems common to theater, opera, music and dance; a twentieth century fund study*. Cambridge, Mass. M.I.T. Press.

Beaud, S., & Weber, F. (2017). *Guide de l'enquête de terrain: Produire et analyser des données ethnographiques*.

Becker, H. S. (1982). *Art worlds*. Berkeley, University of California Press.

Bechky, B. (2006). Gafers, gofers, and grips: Role-Based coordination in temporary organizations. *Organization Science*, 17(1), 3–21.

Bendell, J. (2018). Deep adaptation: A map for navigating climate tragedy. *Insitute for Leadership and Sustainability Occasional Papers*, 2, 1–31.

Bernhardt, A. (2001). *Divergent paths: Economic mobility in the new American labor market.* Russell Sage Foundation

Boltanski, L., & Chiapello, È. (2011). *Le nouvel esprit du capitalisme.* Paris: Gallimard.

Borge, L.-E., Hove, K., Lillekvelland, T., & Tovmo, P. (2017). The Baumol Effect in Defence and Public Administration. Norwegian University of Science and Technology Working Paper.

Bosch, G. (2004). Towards a New Standard Employment Relationship in Western Europe. *British Journal of Industrial Relations, 42*(4), 617–636. https://doi.org/10.1111/j.1467-8543.2004.00333.x

Bourdieu, P. (1971). Le marché des biens symboliques. *L'Année Sociologique, 22,* 49–126.

Bourdieu, P. (1979). *La distinction : Critique sociale du jugement.* Paris: Éditions de Minuit.

Brandellero, A., & Pfeffer, K. (2015). Making a scene: Exploring the dimensions of place through Dutch popular music, 1960–2010. *Environment and Planning A: Economy and Space, 47*(7), 1574–1591. https://doi.org/10.1177/0308518X15595781

Braun, V., & Clarke, V. (2006). Using thematic analysis in psychology. *Qualitative Research in Psychology,* 3(2), 77–101. https://doi.org/10.1191/1478088706qp063oa

Broadberry, S. N., & Crafts, N. F. R. (2009). European productivity in the twentieth century: Introduction. *Oxford Bulletin of Economics and Statistics, 52*(4), 331–341. https://doi.org/10.1111/j.1468-0084.1990.mp52004001.x

Burt, R. S. (2004). Structural Holes and Good Ideas. *American Journal of Sociology,* 110(2), 349–399. https://doi.org/10.1086/421787

Callon, M. (1986). Éléments pour une sociologie de la traduction: la domestication des coquilles Saint-Jacques et des marins-pêcheurs dans la baie de Saint-Brieuc. *L'Année Sociologique,* 36, 169–208.

Cattani, G., & Ferriani, S. (2008). A Core/Periphery Perspective on Individual Creative Performance: Social Networks and Cinematic Achievements in the Hollywood Film Industry. *Organization Science,* 19(6), 824–844. https://doi.org/10.1287/orsc.1070.0350

Chourmouziadou, K., & Kang, J. (2008). Acoustic evolution of ancient Greek and Roman theatres. *Applied Acoustics, 69*(6), 514–529. https://doi.org/10.1016/j.apacoust.2006.12.009

Collins, R. (2004). *Interaction ritual chains.* Princeton, N.J: Princeton University Press.

Collins, R. (2013). Entering and leaving the tunnel of violence: Micro-sociological dynamics of emotional entrainment in violent interactions. *Current Sociology, 61*(2), 132–151. https://doi.org/10.1177/0011392112456500

Corsani, A. (2012). Autonomie et hétéronomie dans les marges du salariat: Les journalistes pigistes et les intermittents du spectacle porteurs de projets. *Sociologie du travail, 54*(4), 495–510. https://doi.org/10.4000/sdt.2122

Cottingham, M. D. (2012). Interaction Ritual Theory and Sports Fans: Emotion, Symbols, and Solidarity. *Sociology of Sport Journal, 29*(2), 168–185. https://doi.org/10.1123/ssj.29.2.168

Cowen, T. (1996). Why I don't believe in the cost disease. *Journal of Cultural Economics,* 20(3), 207–214.

Crossley, N. (2015). *Networks of sound, style and subversion: the punk and post-punk worlds of Manchester, London, Liverpool and Sheffield, 1975–1980.* Manchester University Press. https://doi.org/10.7765/9781847799937

DeNora, T. (2000). *Music in everyday life.* Cambridge, Cambridge University Press.

DiMaggio, P. J., & Powell, W. W. (1983). The Iron Cage Revisited: Institutional Isomorphism and Collective Rationality in Organizational Fields. *American Sociological Review, 48*(2), 147. https://doi.org/10.2307/2095101

Dogan, M. (2008). Strategies in comparative sociology. In M. Sasaki (Éd.), *New Frontiers in Comparative Sociology.* BRILL. https://doi.org/10.1163/ej.9789004170346.i-466.10

Domínguez Rubio, F. (2012). The Material Production of the Spiral Jetty: A Study of Culture in the Making. *Cultural Sociology, 6*(2), 143–161.

Domínguez Rubio, F. (2014). Preserving the unpreservable: docile and unruly objects at MoMa. *Theory and Society,* 43(6), 617–645.

Domínguez Rubio, F. (2016). On the discrepancy between objects and things: An ecological approach. *Journal of Material Culture, 21*(1), 59–86.

Dowd, T. J., & Pinheiro, D. L. (2013). The Ties Among the Notes: The Social Capital of Jazz Musicians in Three Metro Areas. *Work and Occupations,* 40(4), 431–464. https://doi.org/10.1177/0730888413504099

Dubois, J., Durand, P., & Winkin, Y. (2013). Aspects du symbolique dans la sociologie de Pierre Bourdieu. *COnTEXTES.* https://journals.openedition.org/contextes/5661

Durkheim, E. (1912). *Les formes élémentaires de la vie religieuse: Le système totémique en Australie.* J.-M. Tremblay. https://doi.org/10.1522/cla.due.for2. Paris: Quadrige

Elder-Vass, D. (2008). Searching for realism, structure and agency in Actor Network Theory. *The British Journal of Sociology,* 59(3), 455–473.

Elder-Vass, D. (2015). Disassembling Actor-network Theory. *Philosophy of the Social Sciences,* 45(1), 100–121.

Fast, K., Örnebring, H., & Karlsson, M. (2016). Metaphors of free labor: A typology of unpaid work in the media sector. *Media, Culture & Society,* 38(7), 963–978. https://doi.org/10.1177/0163443716635861

Faulkner, R. R., & Becker, H. S. (2009). *Do you know? The jazz repertoire in action.* Chicago: The University of Chicago Press.

FEDELIMA. (2016). *Observation Participative et Partagée.*

Felton, M. V. (1994). Evidence of the existence of the cost disease in the performing arts. *Journal of Cultural Economics,* 18(4), 301–312. https://doi.org/10.1007/BF01079761

Flanagan, R. J. (2012). *The perilous life of symphony orchestras: Artistic triumphs and economic challenges.* New Haven: Yale University Press.

Flocco, G., & Vallée, R. (2012). Une sociologie visuelle du travail: Filmer les machinistes du cinéma et de l'audiovisuel. *Ethnographiques.org,* (25). https://www.ethnographiques.org/2012/Flocco-Vallee

Florida, R. L. (2005). *Cities and the creative class.* Routledge. https://www.myilibrary.com?id=11364

Franssen, T., & Kuipers, G. (2013). Coping with uncertainty, abundance and strife: Decision-making processes of Dutch acquisition editors in the global market for translations. *Poetics,* 41(1), 48–74. https://doi.org/10.1016/j.poetic.2012.11.001

Frey, B. S. (1994). The economics of music festivals. *Journal of Cultural Economics,* 18(1), 29–39. https://doi.org/10.1007/BF01207151

Gomart, E., & Hennion, A. (1999). A Sociology of Attachment: Music Amateurs, Drug Users. *The Sociological Review,* 47, 220–247.

Granovetter, M. (1985). Economic Action and Social Structure: The Problem of Embeddedness. *American Journal of Sociology,* 91(3), 481–510.

Greer, I. (2016). Welfare reform, precarity and the re-commodification of labour. *Work, Employment and Society*, 30(1), 162–173. https://doi.org/10.1177/0950017015572578

Grégoire, M. (2012). Le plein-emploi comme seule alternative à la précarité?: Les intermittents du spectacle et leurs luttes (1919–2003). *Savoir/Agir, 21*(3), 29. https://doi.org/10.3917/sava.021.0029

Grégoire, M. (2013). Les intermittents du spectacle: Le revenu inconditionnel au regard d'une expérience de socialisation du salaire. *Mouvements, 73*(1), 97. https://doi.org/10.3917/mouv.073.0097

Grignon, C., & Passeron, J. C. (1989). *Le savant et le populaire : Misérabilisme et populisme en sociologie et en littérature*. Paris: Gallimard: Seuil.

Griswold, W. (2008). *Cultures and societies in a changing world* (3rd ed). Pine Forge Press.

Griswold, W., Mangione, G., & McDonnell, T. E. (2013). Objects, Words, and Bodies in Space: Bringing Materiality into Cultural Analysis. *Qualitative Sociology*, 36(4), 343–364.

Guibert, G. (2006). *La production de la culture: Le cas des musiques amplifiées en France; genèse, structurations, industries, alternatives*. Séteun.

Hamersveld, I. (Ed.). (2009). *Cultural policy in the Netherlands (Edition 2009)*. The Hague: Amsterdam: Ministry of Education, Culture, and Science; Boekmanstudies.

Hebdige, D. (1991). *Subculture: The meaning of style*. London; New York: Routledge.

Hedegard, D. (2015). Transnational connections: The meaning of global culture in the tastes of Brazilian elites. *Poetics, 53*, 52–64. https://doi.org/10.1016/j.poetic.2015.08.001

Heider, A., & Warner, R. S. (2010). Bodies in Sync: Interaction Ritual Theory Applied to Sacred Harp Singing. *Sociology of Religion, 71*(1), 76–97. https://doi.org/10.1093/socrel/srq001

Hennion, A. (1981). *Les professionnels du disque : Une sociologie des variétés*. A.M. Métailié.

Hennion, A. (2007). *La passion musicale: une sociologie de la médiation*. Paris, Métailié.

Hesmondhalgh, D. (1997). The cultural politics of dance music. *Soundings*, 5, 167–178.

Hesmondhalgh, D. (2006). Bourdieu, the media and cultural production. *Media, Culture & Society, 28*(2), 211–231. https://doi.org/10.1177/0163443706061682

Hesmondhalgh, D., & Baker, S. (2010). 'A very complicated version of freedom': Conditions and experiences of creative labour in three cultural industries. *Poetics*, 38(1), 4–20. https://doi.org/10.1016/j.poetic.2009.10.001

Hitters, E., & Richards, G. (2002). The creation and management of cultural clusters. *Creativity and innovation management*, 11(4), 234–247.

Hitters, E., & van de Kamp, M. (2010). Tune in, fade out: Music companies and the classification of domestic music products in the Netherlands. *Poetics*, 38(5), 461–480.

Horning, S. S. (2004). Engineering the Performance: Recording Engineers, Tacit Knowledge and the Art of Controlling Sound. *Social Studies of Science*, 34(5), 703–731.

Horning, S. S. (2013). *Chasing sound: Technology, culture, and the art of studio recording from Edison to the LP* (Johns Hopkins University Press).

Hughes, E. C., & Chapoulie, J.-M. (1996). *Le regard sociologique: Essais choisis*. Paris: Éd. de l'École des Hautes Études en Sciences Sociales.

Jeanpierre, L., & Roueff, O. (Éd.). (2014). *La culture et ses intermédiaires : Dans les arts, le numérique et les industries créatives*. Paris (France) : Éditions des archives contemporaines.

Jessop, B. (2012). Economic and Ecological Crises: Green new deals and no-growth economies. *Development, 55*(1), 17–24. https://doi.org/10.1057/dev.2011.104

Jiménez Sedano, L. (2019). "From Angola to the world", from the world to Lisbon and Paris: How structural inequalities shaped the global kizomba dance industry. *Poetics*, *75*, 101360. https://doi.org/10.1016/j.poetic.2019.04.001

Johnson, C., Dowd, T. J., & Ridgeway, C. L. (2006). Legitimacy as a Social Process. *Annual Review of Sociology*, *32*(1), 53–78. https://doi.org/10.1146/annurev.soc.32.061 604.123101

Kaplan, D. (2012). Institutionalized erasures: How global structures acquire national meanings in Israeli popular music. *Poetics*, *40*(3), 217–236. https://doi.org/10.1016/j.poetic.2012. 03.001

Kealy, E. R. (1979). From Craft to Art: The Case of Sound Mixers and Popular Music. *Sociology of Work and Occupations*, *6*(1), 3–29.

Klett, J. (2014). Sound on Sound: Situating Interaction in Sonic Object Settings. *Sociological Theory*, *32*(2), 147–161.

Kuipers, G. (2011). Cultural Globalization as the Emergence of a Transnational Cultural Field: Transnational Television and National Media Landscapes in Four European Countries. *American Behavioral Scientist*, *55*(5), 541–557. https://doi.org/10.1177/000276421139 8078

Kuipers, G. (2015). How National Institutions Mediate the Global: Screen Translation, Institutional Interdependencies, and the Production of National Difference in Four European Countries. *American Sociological Review*, *80*(5), 985–1013.

Laborde, D. (2008). L'Opéra et son régisseur: Notes sur la création d'une œuvre de Steve Reich. *Ethnologie française*, *38*(1), 119. https://doi.org/10.3917/ethn.081.0119

Lamont, M., & Molnár, V. (2002). The Study of Boundaries in the Social Sciences. *Annual Review of Sociology*, *28*(1), 167–195.

Latour, B. (1990). Technology is Society Made Durable. *The Sociological Review*, *38*, 103–131. https://doi.org/10.1111/j.1467-954X.1990.tb03350.x

Latour, B. (2007). *Reassembling the social: an introduction to Actor-Network-Theory*. Oxford, Oxford Univ. Press.

Law, J. (2009). Actor Network Theory and Material Semiotics. In B. S. Turner (Dir.), *The New Blackwell Companion to Social Theory*. Oxford, UK: Wiley-Blackwell: 141–158

Le Guern, P. (2004). Mutations techniques et division du travail: le cas des monteurs sons. *Volume!*, *3*(0).

Lena, J. C. (2012). *Banding together: How communities create genres in popular music*. Princeton; Oxford: Princeton University Press.

Lena, J. C. (2015). Relational approaches to the sociology of music. In L. Hanquinet & M. Savage (Éd.), *Routledge International Handbook of the Sociology of Art and Culture*. Routledge.

Lettow, S. (2017). Turning the turn: New materialism, historical materialism and critical theory. *Thesis Eleven*, *140*(1), 106–121.

Leyshon, A. (2009). The Software Slump?: Digital Music, the Democratisation of Technology, and the Decline of the Recording Studio Sector within the Musical Economy. *Environment and Planning A*, *41*(6), 1309–1331.

Liebst, L. (2019). Exploring the Sources of Collective Effervescence: A Multilevel Study. *Sociological Science*, *6*, 27–42. https://doi.org/10.15195/v6.a2

Lingo, E. L., & O'Mahony, S. (2010). Nexus Work: Brokerage on Creative Projects. *Administrative Science Quarterly*, *55*(1), 47–81.

Lizé, W. (2016a). Artistic work intermediaries as value producers. Agents, managers, tourneurs and the acquisition of symbolic capital in popular music. *Poetics*, 59, 35–49.

Lizé, W. (2016b). La légitimité du jazz et des musiques savantes : Des statistiques sur les publics à la critique en ligne. *RESET*, (5). https://doi.org/10.4000/reset.622

Maguire, J. S., & Matthews, J. (2012). Are we all cultural intermediaries now? An introduction to cultural intermediaries in context. *European Journal of Cultural Studies*, 15(5), 551–562.

Maisonneuve, S. (2009). *L'invention du disque 1877–1949: Genèse de l'usage des médias musicaux contemporains*. Archives contemporaines.

Mauss, M. (1936). Les techniques du corps. *Journal de Psychologie*, 32(3–4), 271–293.

McCormick, L. (2006). Music as a social performance. In R. Eyerman & L. McCormick, *Myth, Meaning and Performance*. New York: Routledge

McGrath, T., Legoux, R., & Sénécal, S. (2017). Balancing the score: The financial impact of resource dependence on symphony orchestras. *Journal of Cultural Economics*, 41(4), 421–439. https://doi.org/10.1007/s10824-016-9271-z

McLeod, K. (2001). Genres, Subgenres, Sub-Subgenres and More: Musical and Social Differentiation Within Electronic/Dance Music Communities. *Journal of Popular Music Studies*, 13(1), 59–75.

Meerkerk, E. van, & Hoogen, Q. van den (Eds.). (2018). *Cultural policy in the polder: 25 years Dutch Cultural Policy Act*. Amsterdam: Amsterdam University Press.

Menger, P.-M. (1999). ARTISTIC LABOR MARKETS AND CAREERS. *Annual Review of Sociology*, 25(1), 541–574. https://doi.org/10.1146/annurev.soc.25.1.541

Menger, P.-M. (2015). *Les intermittents du spectacle : Sociologie du travail flexible* (Nouv. éd). Editions de l'Ecole des hautes études en sciences sociales.

Mercier D. (Dir.), (2017). *Le livre des techniques du son. Tome 2 – La technologie*. Paris, Dunod.

Metiu, A., & Rothbard, N. P. (2013). Task Bubbles, Artifacts, Shared Emotion, and Mutual Focus of Attention: A Comparative Study of the Microprocesses of Group Engagement. *Organization Science*, 24(2), 455–475. https://doi.org/10.1287/orsc.1120.0738

Montanari, F., Scapolan, A., & Gianecchini, M. (2016). 'Absolutely free'? The role of relational work in sustaining artistic innovation. *Organization Studies*, 37(6), 797–821. https://doi.org/10.1177/0170840616647419

Negus, K. (2002). The work of cultural intermediaries and the enduring distance between production and consumption. *Cultural Studies*, 16(4), 501–515.

Nowell, L. S., Norris, J. M., White, D. E., & Moules, N. J. (2017). Thematic Analysis: Striving to Meet the Trustworthiness Criteria. *International Journal of Qualitative Methods*, 16(1), 160940691773384. https://doi.org/10.1177/1609406917733847

Patriotta, G., & Hirsch, P. M. (2016). Mainstreaming Innovation in Art Worlds: Cooperative links, conventions and amphibious artists. *Organization Studies*, 37(6), 867–887. https://doi.org/10.1177/0170840615622062

Pavis, P. (1988). From text to performance. In M. Issacharoff & R. F. Jones, *Performing texts* University of Pennsylvania Press. Philadelphia. 86–100

Percival, N., & Hesmondhalgh, D. (2014). Unpaid work in the UK television and film industries: Resistance and changing attitudes. *European Journal of Communication*, 29(2), 188–203. https://doi.org/10.1177/0267323113516726

Perrenoud, M. (2007). *Les musicos: enquête sur des musiciens ordinaires*. Paris, Découverte.

Perrenoud, M., & Bataille, P. (2017). Being a Music Performer in French Speaking Switzerland: Relationships to Work and Employment. *Swiss Journal of Sociology*, 43(2), 309–334. https://doi.org/10.1515/sjs-2017-0017

Perrin, T., Delvainquière, J.-C., & Guy, J.-M. (2017). *Country profile: France.* Compendium of cultural policies and trends in Europe/ERICarts. https://www.culturalpolicies.net/dat abase/search-by-country/country-profile/?id=13

Peterson, R. A., & Kern, R. M. (1996). "Changing Highbrow Taste: From Snob to Omnivore". *American Sociological Review*, 61(5), 900.

Peterson, R. A., & Anand, N. (2004). The Production of Culture Perspective. *Annual Review of Sociology, 30*(1), 311–334. https://doi.org/10.1146/annurev.soc.30.012703.110557

Poirrier, P. (Éd.). (2002). *Les politiques culturelles en France.* Paris: Documentation française.

Ragin, C. C. (2008). Measurement Versus Calibration: A Set-Theoretic Approach. In JBox-Steffensmeier, J.M., Brady, H.E., Collier, D. (Eds.), T*he Oxford Handbook of Political methodology,* Oxford University Press, https://doi.org/10.1093/oxfordhb/978019928 6546.003.0008

Reilly, P. (2017). The Layers of a Clown: Career Development in Cultural Production Industries. *Academy of Management Discoveries*, 3(2), 145–164. https://doi.org/10.5465/amd. 2015.0160

Rihoux, B., & Ragin, C. C. (Éd.). (2009). *Configurational comparative methods: Qualitative comparative analysis (QCA) and related techniques.* Thousand Oaks: Sage.

Roberge, J. (2009). Jeffrey C. Alexander et les dix ans du programme fort en sociologie culturelle. *Cahiers de recherche sociologique, 47*, 47. https://doi.org/10.7202/1004979ar

Rudent, C. (2008). Le premier album de Mademoiselle K : Entre création individuelle et coopérations négociées. *Ethnologie française*, 38(1), 69.

Sapiro, G. (2007). La vocation artistique entre don et don de soi. *Actes de la recherche en sciences sociales*, 168(3), 4. https://doi.org/10.3917/arss.168.0004

Schechner, R. (1974). From Ritual to Theatre and Back: The Structure/Process of the Efficacy-Entertainment Dyad. *Educational Theatre Journal, 26*(4), 455. https://doi.org/10.2307/ 3206608

Schechner, R. (1985). *Between Theater and Anthropology:* University of Pennsylvania Press. https://doi.org/10.9783/9780812200928

Schechner, R., & Schuman, M. (Éd.). (1976). *Ritual, play, and performance: Readings in the social sciences/theatre.* Seabury Press.

Schneider, C. Q., & Wagemann, C. (2012). *Set-theoretic methods for the social sciences: A guide to qualitative comparative analysis.* Cambridge: Cambridge University Press.

Schmutz, V. (2009). Social and symbolic boundaries in newspaper coverage of music, 1955–2005: Gender and genre in the US, France, Germany, and the Netherlands. *Poetics*, 37(4), 298–314.

Shapin, S. (1989). The invisible technician. *American Scientist*, 77(6), 554–563.

Singer, B. (2018). *Bohemian Rhapsody* [Biopic; Color – 2,39:1].

Siracusa, J. (2000). Le montage de l'information télévisée. *Actes de la recherche en sciences sociales*, 131(1), 92–106. https://doi.org/10.3406/arss.2000.2668

Sorignet, P.-E. (2010). *Danser: Enquête dans les coulisses d'une vocation.* Paris: Découverte.

Summers-Effler, E. (2002). The Micro Potential for Social Change : Emotion, Consciousness, and Social Movement Formation. *Sociological Theory, 20*(1), 41–60. https://doi.org/10. 1111/1467-9558.00150

Throsby, D., & Zednik, A. (2011). Multiple job-holding and artistic careers: Some empirical evidence. *Cultural Trends*, 20(1), 9–24. https://doi.org/10.1080/09548963.2011.540809

Tournès, L. (2008). *Du phonographe au MP3 : Une histoire de la musique enregistrée, XIXe–XXIe siècle*. Paris: Autrement.

Turner, V. (1982). *From ritual to theatre: The human seriousness of play*. Performing Arts Journal Publications.

Turner, V., & Schechner, R. (1995). *The anthropology of performance* (Nachdr.). PAJ Publ.

Van Der Leden, J. (2017). *Country profile: The Netherlands*. Compendium of cultural policies and trends in Europe/ERICarts. https://www.culturalpolicies.net/database/search-by-country/country-profile/?id=28

van Eijck, K. (2001). Social Differentiation in Musical Taste Patterns. *Social Forces*, 79(3), 1163–1185. https://doi.org/10.1353/sof.2001.0017

van Venrooij, A. (2009). The aesthetic discourse space of popular music: 1985–86 and 2004–05. *Poetics*, 37(4), 315–332.

van Venrooij, A., & Schmutz, V. (2018). Categorical ambiguity in cultural fields: The effects of genre fuzziness in popular music. *Poetics*, 66, 1–18.

Vlegels, J., & Lievens, J. (2017). Music classification, genres, and taste patterns: A ground-up network analysis on the clustering of artist preferences. *Poetics*, 60, 76–89.

VNPF. (2016). *Annual Survey of Members*.

Vonk, G., & Jansen, A. (2017). Social protection of marginal part-time, self-employment and secondary jobs in the Netherlands. *WSI Study*, 9.

Uzzi, B., & Spiro, J. (2005). Collaboration and Creativity: The Small World Problem. *American Journal of Sociology*, 111(2), 447–504. https://doi.org/10.1086/432782

Watson, A. (2014). *Cultural production in and beyond the recording studio* (Routledge). https://www.vlebooks.com/vleweb/product/openreader?id=none&isbn=9781135006310

Wright, D. (2005). Mediating production and consumption: Cultural capital and «cultural workers». *The British Journal of Sociology*, 56(1), 105–121. https://doi.org/10.1111/j.1468-4446.2005.00049.x

Zukin, S., & DiMaggio, P. J. (1990). Introduction. In S. Zukin & P. J. DiMaggio, *Structures of Capital: The Social Organization of The Economy*, 1–36, Cambridge: Cambridge University Press.

Zweig, K. A. (2016). *Network analysis literacy: A practical approach to the analysis of networks*. Springer.